Information Technology

IN CONSTRUCTION DESIGN

Michael Phiri

Thomas Telford

Published by Thomas Telford Publishing, Thomas Telford Ltd, 1 Heron Quay, London E14 4JD.
URL: http://www.t-telford.co.uk

Distributors for Thomas Telford books are
USA: ASCE Press, 1801 Alexander Bell Drive, Reston, VA 20191-4400
Japan: Maruzen Co. Ltd, Book Department, 3–10 Nihonbashi 2-chome, Chuo-ku, Tokyo 103
Australia: DA Books and Journals, 648 Whitehorse Road, Mitcham 3132, Victoria

First published 1999

A catalogue record for this book is available from the British Library

ISBN: 0 7277 2673 0

© Michael Phiri and Thomas Telford Limited, 1999

This book is published on the understanding that the author is solely responsible for the statements made and opinions expressed in it and that its publication does not necessarily imply that such statements and/or opinions are or reflect the views or opinions of the publishers. While every effort has been made to ensure that the statements made and the opinions expressed in this publication provide a safe and accurate guide, no liability or responsibility can be accepted in this respect by the author or publishers.

Designed by Kneath Associates
Printed and bound in Great Britain by Bookcraft (Bath) Ltd

Preface

The tremendous speed and enormous diversity of technological developments often means that many information technology (IT) publications are out of date before they even go to proof. Consequently, opportunities are lost to gather and to structure material, and to learn from experience. This is the main challenge for this book which includes looking at the dynamic IT industry — one full of hype and in constant flux — while seeking to establish its usefulness within the construction industry. The construction industry is, in comparison with IT, an industry noted for its slow rate of change and for its fragmentation, and this effect on the industry's productivity is staggering compared with other business sectors.

Most of us are acutely aware of the tremendous strides made in manufacturing during the past two decades to improve quality and costs, and to implement new technologies. Most of us are also aware that little has changed in the construction industry over the same period. The current state of fragmentation needs to change. The specialization that was so beneficial to the post-war building has degenerated into fragmentation characterized by typical limitations of company interests, adversarial professional protectionism, sequential and discrete activities, and disjointed design processes. Change is needed to create an environment that fosters innovation, clarifies communication and truly integrates the design, manufacturing and construction processes. IT technologies and knowledge-rich environments offer real opportunities and the potential to realize this integration and offer an arrangement in which there is a single contracting entity from which participants provide their expertise. Such a framework promises reduced complexity and seamless integration among multi-disciplines involved in the construction projects (architects, structural engineers, civil engineers, mechanical and electrical engineers, building services engineers, project managers, planners, surveyors, etc.), and it enables systems development to allow sharing of knowledge within organizations, and also externally with other organizations, and allows the merging of skills to deliver the integrated project using the single model for the entire life of a project (pre-project, project and post-project stages) to meet clients' needs and expectations.

Furthermore, having IT policies and tools does not mean abandoning conventional skills and practices. Having sound IT policies and knowledge of hardware platforms, operating systems and software, alongside traditional tools, emphasizes the applications over the transient consumer products and, overall, makes good business sense by promising to provide an individual organization with choices as well as a competitive edge of strategic advantage. An equally important point to remember is that having IT policies and knowledge requires the user to distinguish the following three management decision-making and operational levels.

1   *Strategic* — that is providing a long-term framework; a plan defining the future commitment of an organization's resources, often over a period of ten or more years, and using information that is low in volume but broad in scope in order to make the long-term strategic decisions with wide-ranging effects.

2   *Tactical* — that is providing controlling mechanisms which compare results of operations with plans and adjusting these accordingly, often over a period of about five years and involving summaries and interpreted information.

3   *Operational* — that is providing means to facilitate the day-to-day flow of work ('fire-fighting' and reactive work) and involving high-volume, detailed data of everyday activities and weekly trends.

Also, many organizations often concentrate on the selection and the purchase while neglecting the application of IT systems. Selecting and purchasing software and hardware is only a first step, further steps have to be taken to ensure successful application and implementation. For example, needs have to be defined in business terms to identify measurable and verifiable goals (such as cost reductions and efficient operations), and this may include turnaround times, quality improvements, monitoring performance, tracking equipment, providing timely information, reducing energy costs on heating and ventilation by 20%, HVAC efficiency, equipment safety, etc. The application of specific measures is just as important as the prioritization of objectives and the setting-up of a realistic timescale for their achievement. Also, purchasing hardware, for instance with a suitably large memory and sophisticated software, such as visualization, is a big investment not to be taken lightly and once made it then requires effective, intensive use and training to ensure that staff have the necessary skills in order to realize and maximize the return on the investment.

As the title suggests, this book is aimed foremost at the related professions of architects, engineers, contractors and subcontractors, facilities and resources managers in construction; in other words, all those individuals involved in the design and maintenance of the built environment. Often there is a perception that each of these groups has different organizational, strategic and tactical problems. The main intention is to build bridges between disciplines by seeking to write in general, rather than in specialist, terms. The approach is holistic and needs to bear in mind the breadth of this readership.

The book is also for all those concerned with successful development and implementation of construction procedures and methodologies. Users and managers of information technology, students, trainees, draftspersons and others in construction and the related professions will find the book an invaluable resource guide to how different organizations or practices use IT in the context of construction.

The book consists of a number of themes. The first theme concerns an emphasis on the historical perspective and context. The main intention is the acquisition of knowledge about the development of IT tools in order to give insights which might be invaluable for their use and implementation. Learning from the experience of other sectors is important for the future briefing and procurement strategies of construction clients with ongoing building programmes, and also in identifying trends, themes, drivers and the main issues concerned with technology applications. Some of the problems are not unique to the construction industry.

The second theme concerns bridging the gap between practice and theoretical developments of computer software and hardware tools while focusing on the realities of professional architectural/engineering practice. These developments and resources are often presented in trade literature and publications designed for specific users, for example computer professionals (*Computer World, Compute!*); for programming and operating systems users (*Dr Dobb's Journal, User's Guide for MS-DOS*); for

software writers and specifiers (*Software Digest, Personal Software*). These numerous magazines, including *Byte, PC World, Macworld, PC Magazine, Macuser, In Cider, Nibble, Publish,* etc., are for enthusiasts or 'nerds' but are not readily available, applicable or meaningful for busy practitioners concerned more with meeting project deadlines and dealing with immediate problems as opposed to formulating and implementing IT strategies and policies. In order to allow easy access, the book seeks to bring together, as much as is practical in a single volume, all this invaluable information.

Furthermore, the concern with the design process, algorithms, methodologies and objectives means that the preferred method of illustration in the book is the process flow-chart rather than screen shots, which are often popular in most books of this type.

The third theme concerns reference to IT in action, notably focusing on the experience of a number of UK case study organizations (from different markets) rather than individual construction projects. Each organization has a story to tell and this story includes a wealth of experience from past and present projects, obstacles which hinder the design process, associated success and failures and all of which are worth sharing. Often practices are unaware of how others in their field operate and this publication attempts to provide an accessible and route-finding medium to share, observe and disseminate ideas and ways in which other practices approach the subject matter. This is not to say that this book is an authoritative technology treatise for IT for architects and engineers.

The fourth theme is an emphasis on IT applications. For the case studies this means looking beyond the project delivery process as a one-off event or viewing each project as a prototype with each problem, site or client being perceived as unique, to considering the entire building-cycle involving a series of projects in a continuous building programme. There is widespread evidence that the majority of existing project processes are unnecessarily complex, haphazard, unsystematic and inefficient leading to abortive design time, construction and management costs, and to products and results which do not fulfil the needs and expectations of building clients. Successful IT applications offer opportunities of cost reduction and means of improving the quality, efficiency and cost of their building programmes by means of harnessing information technology into standard briefing and specification processes in all sectors (retailers, corporates and public sector) with frequent investment in construction.

The fifth and underlying theme is also the main reason for writing this book, that is the need to contribute towards an increased understanding and systematization of the design process through a focus on its constituent parts in order to enhance design effectiveness. My view is that the rationalization of the briefing and evaluation process to achieve cost and time savings and thereby improve overall performance is facilitated by information technology. Both the processes and their standardization are not well researched and remain effectively unmeasured or misunderstood. There continues to be a misunderstanding of their impact on the briefing of the processes (standardization), on policies (standards) and on standardization of products (standard parts). Equally, there is lack of factual data on the resistance to standardization and, particularly, why the standardization of processes has not been more readily acceptable. Current research in universities and in private and public organizations, such as traditional construction, has focused mainly on the use of standard products, and glossed over standardizing the process. Against this background, the benefits of rationalization of the client briefing and the evaluation process with a sound IT basis are obvious. For example, in the automotive manufacturing industry, time compression, late configuration, specialization, empowerment as well as standardization have led to significant step-change improvements in the value that clients obtain from projects.

The sixth theme concerns the approach taken throughout this book, namely that of celebrating and glorifying IT as the wonderful new complex technology which many people do not fully understand yet but must still possess. This is not to play down the psychological and social consequences and possible mishaps, the so-called normal accidents, disasters or even catastrophes which can be associated with these new all-encompassing technological developments. All of which require planning, regular review of policies and procedures, for instance recovery from mishaps such as loss of data, problems with consumables, problems with hardware at a critical time in a project process, etc. In trying to use computers to do all things in life that are tiring, repetitive and monotonous or by allowing computers to take over the drudgery while we concentrate on matters that are amusing and interesting we may not be doing humanity any favour.

Acknowledgements

This book includes contributions by many architectural and engineering practices who form the case study organizations. I express my gratitude to the individuals of these practices and organizations whose help was invaluable and without whom the manuscript would not have been possible: Peter Angrave, Joselyn Van De Bosche, Phillip Ball, Richard Crawford, Colin Davis, Charles Davis, Nigel Davis, Paul Davis, Ian Davidson, Nick Dunn, Rhodri Evans, Ross Gates, John Gill, Neville Granville, Piers Heath, Lars Hesselgren, Linda Harbour, Harry McCourt, Neil Mitchel, Robert Myers, Terry Nichols, Martin Perry, Chris Poulton, Mike Purvis, Leandro Rotondi, Dr T. Selman, Paul Slaney, John Thompson, Stuart Small, Fiona Charlesworth, Simon Williams-Gunn, Derek Winsor and Richard Worthington. I am grateful to them for giving me their time.

My thanks to everyone, especially Daker Fleming, who was involved in refereeing the book and for the useful feedback.

For their contagious enthusiasm, many thanks go to the Briefing & Monitoring Research Team at the Institute of Advanced Architectural Studies, University of York — Professor John Worthington, Adrian Leaman, Dr Rama Isiah and Sally Kirk-Walker; and to Anne Dicks, Secretary at the Institute. I wish to extend this gratitude to Professor Bryan Lawson at the School of Architecture, University of Sheffield where the Building Performance and Design Brief Management Research is now continuing from the Institute of Advanced Architectural Studies.

Also, I would like to thank the editors and all those who contributed to the mark-up of the manuscript and whatever else was involved in the production of the book, and in particular to Victoria Wheeler, Alex Lazarou and Steve Temblett of Thomas Telford Publishing, and Val Kinsler and others for their hard work, trust, patience and encouragement.

Finally, my special acknowledgement and dedication goes to the support, tolerance, understanding and encouragement of my mother and father, Annabel, Tobias, Oscar, Selina, Phoebe, Violet, Faneri and Richard over what seemed eternity.

Permissions

Part 1

Figures 3, 4(a) and 6 reproduced from *The First Electronic Computer* by Alice Burks and Arthur Banks, © The University of Michigan Press, 1988

Figure 4(b) reproduced from *From ENIAC to UNIVAC: an appraisal of Mauchly Computers* by Nancy Stern, © Digital Press/Compaq Computer Corporation and Butterworth-Heinemann/Reed Elsevier, 1981

Figures 12(a–b) © Professor M. Saleh Uddin, School of Architecture, Southern University, Louisiana, USA

Figures 13(a–b) and 14(a–j) © Asta Development Corporation Ltd, Oxfordshire, UK, 1998

Figure 18 © Atlantic EC Ltd, Bradford, UK, 1998

Figure 20 reproduced from *Architecture Today*, 21 Sept. 1991, © Westward Ltd, 1998

Figures 21(a–d), 22(a–c), 25(a–c) and 26 reproduced courtesy of NHS Estates, © Crown copyright 02/11/1998

Figure 24 reproduced from *Principles of computer-aided design* edited by Rooney and Steadman, © UCL Press/Open University Press and courtesy of rights manager Taylor and Francis Ltd

Figure 27(a–b) reproduced courtesy of Oxford Method of Building, Department of Health © Crown copyright

Figure 28 reproduced courtesy and © of Gable CAD System Ltd and Price Waterhouse Coopers

Contents

and at IT in action. Finally, Part 3 provides some comments, examining emerging views and future developments.

Following the introduction, Part 1 begins with a background look at the computer system as a collection of parts (input, system and output) and examines the milestones in the history of computing. The next four chapters cover text and graphics manipulation including specification writing, project management, financial accounting/book-keeping and modelling and database activities respectively.

The important point is that IT can save time and make management more effective, thus giving us better information. IT is the key information knowledge facilitator. One observation is that IT is 'pervading virtually all forms of human endeavour'; work, education and leisure, communication, production and marketing, and the time scheduling of these.

IT is changing the scale and content of information networks, the interdependence of organizations and how as well as where we live, work, shop, learn, communicate and play.[2]

IT underpins virtually every aspect of the modern business and can make all the difference to an organization's competitiveness. The merger of data, computing and communications has invaded virtually every aspect of social life. Computers, originally intended as fast calculators, are now recognized as the primary tool with which to access, understand and disseminate vast quantities of information through images, text and sound, via the media of television, telephone, compact disks (CDs) and the Internet. This democratizing of information makes it accessible and understand-able to a greater number of people. IT enhances, to a very significant extent, the number of people that can engage in the organization of knowledge or expertise, as well as the speed and the volume with which this can be processed and exchanged — the dynamics and diversification of economic activity. Yet it is worth remembering that IT is only an aid to management, not a facsimile substitute, and furthermore print-outs are not management, neither is electronic mail effective supervision.

Nonetheless, the computer will not change us but will help us to innovate by taking away some of the repetitive tasks; doing what architects/engineers already do but faster and with more accuracy. This means allowing the addition of testing to let us see more clearly where we can innovate.

The chapter on data manipulation and statistical analyses is equally important. It describes readily available appropriate tools for both architects and engineers, which might otherwise remain unused because of a failure to understand their relevance. The final two chapters of Part 1 cover CAD/CAM/CAE, and telecommunications and networks.

Various design aspects are looked at, starting from the briefing stage to completion of a project. At each stage, the way information is handled is considered, including the impact of recent technologies, with reference to case studies in architectural and engineering practice. For instance, with little or no imagination even production information for particular projects can be made clearer and more pleasing; colour becomes as easy to use as monochrome even for multiple print drawings, thereby making the comeback of the beauty of coloured Victorian working drawings possible.

Survey photographs can also be put on CD and then included as pictures on drawings, or assembly information from manufacturers' literature can be scanned and pasted in specifications or drawings, giving idiot-proof information in graphic form next to relevant technical details. Potential production information can be given which fully exploits the ability to combine, mix and store drawing, text, graphic and spreadsheet data easily in an integrated way, thus breaking down the usual distinction between written specification and drawing details.

In the area of conceptual presentations, the potential for manipu-lating data and images in on-screen documents adds a new dimension to the drama of architectural presentation. The possibility now exists of using computer models to present architectural projects, using the seamless graphics and animations commonplace in TV and cinema. These can vary from simple preset slide shows to sophisticated interactive documents consisting of a number of pages containing images imported from other documents, using a variety of 2D drawings, photographs, scanned pictures, text, pictures and animations of 3D models, video clips of the site as it exists, and more. The ability to interweave drawing and all kinds of such presen-tations offers advantages over hard copy drawings and models, including ease of assembly, ease of transport, iconization via screen, ease of alteration, and duplication or transfer to video and, of course, no requirements to insure models. Renderings, 'walkthrough' and animations help architects/engineers compete for projects by making their proposals stand out from the others.

For any investment, a business needs a return in terms of increased revenue or reduced costs. Savings in time created by CAD systems can be direct cost savings or a revenue opportunity, depending on what you do with it. The search for opportunities to increase productivity, quality, serviceability and competitiveness is a priority for more and more users.

The goal of the architect/engineer in the Information Age is not necessarily to compress the time required to produce traditional documentation, but to exploit the amount and nature of information available about a proposed building/facility to the benefit of the building's designers, users and owners. The role of computers and their sophisticated CAD software not only allows realization of concepts previously thought or deemed to be unrealizable, but also affords a richness previously possible only with very expensive craftsmanship to create monuments to the age of the microchip.

Architects and engineers have access to a wealth of both commercial and non-commercial information, and none of it is in their office. If this information is at the end of a telephone line then the costs of retrieving it are significantly cheaper than maintaining a massive library of perpetually out-of-date technical manuals. A computer can connect you, via a telephone line, to vast libraries in a matter of minutes. Instead of shelves upon shelves of binders, you can download (i.e. to your own computer) only the industry or product specification you need.

The problem is to select the most reliable, accurate, helpful items from the millions of promotional and trivial facts that litter the information superhighway. The key to quick retrieval is knowing where to go and what kind of web site contains what type of information. You can use electronic mail (E-mail) to communicate. It is cheaper and less wasteful than faxes. Easy access to electronic forums and 'chat' rooms permits on-line users to debate issues, get technical questions answered — even place or answer classified ads. For most architects, many of the advantages of on-line information services still lie in the future, but with the low cost of modems and services and the steadily improving quality of information, being on-line will soon become as mandatory for practices as a fax machine.

Using the Internet and the web, which are rapidly evolving to become the infrastructure for the global economy, architects/engineers will eventually be able to search manufacturers' libraries for appropriate products and place these in their project on the computer, complete with 3D product model, specifications and drafting symbols. Whenever the project file is opened, the building products will 'read' that a file has been downloaded from their manufacturer and alert the architect if any specifications have changed.

Drawing on case studies from architectural and engineering firms, Part 2 assesses IT in action, looking at the way practices strategically organize and resource to use IT, at various stages of the project and from project to project and dealing with both project-specific data as well as general or non-project-specific data from catalogues of building materials and components, libraries of standard details and descriptions of previous projects. The firms chosen for detailed study cover

- different sizes: e.g. small, medium, large, very large, etc.

- different structures and ownership patterns: e.g. sole proprietor, private partnership, limited company, etc.

- different services offered to clients in very different markets: e.g. architectural and/or engineering design, design-based consultancy, multi-disciplinary, etc.

- different philosophies and working methodologies.

No attempt is made to suggest that the firms represent a statistically relevant sample, or a comprehensive range of possible types. There are omissions, such as those from the public sector. Nor are they randomly chosen. They were selected on the basis of knowledge of the firms, the respect that they command within the architectural and engineering professions and the likelihood that their histories would shed some light on how IT changes have taken place and have affected their processes. The chosen practices appear to show enthusiasm for the use of computers and new technology in their daily operation — for instance using computer techniques equivalent to drawing and physical model making and as design tools. The case study organizations include

- single-principal practices dependent on the reputation of one person and this individual seems to take a hand in every project and consequently rarely runs one

- federated practices sharing resources and some staff

- larger practices that are partnerships in which the partners all lead for particular jobs or take corporate roles

- practices configured more along corporate lines with each partner playing a particular and defined role, taking responsibilities for certain stages of work rather than for individual projects.

Interviews were undertaken based on a checklist of minimum desirable details relating to each firm. The object was to gain an insight into the working of the practice rather than a focus on the design achievement. Questions sought to discover trends and main issues. The information involved looking at the background of the organizations and IT applications — text manipulation, databases, specification writing; CAD, telecommunications and networks; training and selected projects.

Part 3 provides some comments about the above studies. One conclusion from an examination of these architectural and engineering practices is an acknowledgment of the relevance of IT as a tool to aid the design process. The problems are those affecting the industry as a whole — the pace of change in technology; issues of inter-operability; web-based e-commerce or Internet technology posing new sets of problems of security and speed of access (or lack of it).

There is also a need to reconfigure systems to deal with the millennium bug or to cope with the introduction of the single European currency. Another debate concerns the personal computer and the network computer (NC) — the battle between fat client PCs and thin NCs which give control over to the IT manager and system administrators to reduce the overall running costs of desktop computing devices. The single point of control and complexity allows software on the central server to reduce worries about updating many machines or unauthorized copying of data which carries the risk of infecting the whole system by inserting a disk with a virus.

The telephone system — voice mail, call diversion, remote pick-up and so on, controlled via a switchboard and not the handset — is an example of this idea of single point of complexity and control with a large number of simple devices attached to it.

Most of the developments in IT have involved the PC boom in the 1980s and the rise of client/server systems in the 1990s, all of which have meant exponential progression in speed, storage capacity, data crunching ability and cost-efficiency — i.e. doing more of the same for less money. Recent changes, however, indicate how IT is now being used to do things differently — and to do different things — thereby presenting us with new challenges. Architects/engineers have been drafting for more than 400 years. They have automated their drafting only in the past 20 years but for the next 20 years they will use 3D-based software to simulate buildings. The

simulated buildings will create opportunities for new or enhanced architectural/engineering services such as

- sharing of information
- computer rendering, animation and 'virtual reality' scenes to help community groups, financial backers or prospective tenants or customers visualize the designs in 3D facilities management
- simulation and visualization of buildings material performance
- exploration of design and maintenance alternatives
- feasibility studies of alterations
- simulation and planning of design changes required over the life of buildings
- optimization of energy use.

Increasingly sophisticated structural analysis now available on personal computers is helping engineers to meet the demand for long clear spans for buildings (projects ranging from warehouses, factories and passenger terminals to entertainment complexes) and bridges. The more advanced and flexible computer programs allow full use of limit state design, i.e. each member to be designed down to its most economical size, taking all loading conditions into account (wind, dead and imposed loads). A re-iterative process to include the self-weight of steelwork is performed until an optimum design is reached. For bridges, software developments such as 3D non-linear analysis programs and sheer computing capacity have vastly improved design methods and enabled far more accurate calculation of environmental and other loading effects. The end result is greater structural efficiency, which translates into reduced weight and longer spans.

At the end of the book a comprehensive glossary and list of acronyms are provided for quick and easy reference, in order to demystify and make sense of it all, so that buying and talking about computers is simplified.

References

1. Longley D. and Shain M. *Dictionary of information technology*. Third edn, Macmillan, 1989, 165.

2. Brotchie J. F., Hall P. and Newton P. A. (eds). *The spatial impact of technological change*. London, Croom Helm, 1987, xv.

Background

Introduction

This chapter begins with systems definitions and proceeds to a review of notable landmarks in computing history across its five generations, and concludes that these important stages form the basis for some of the emerging trends. These include

- trends in information technology and telecommunications towards convergence of products to produce multi-functional products (i.e. telecoms equipment incorporating more and more aspects of computing, and computers incorporating more and more aspects of telecoms)

- trends towards miniaturization of electronic products driven by requirements for portability, e.g. downsizing and nanotechnology.[1,2] (Nanotechnology uses devices such as the equivalent of solenoids, pipes and pumps to create microscopic 'factories' of assemblers and disassemblers. These diminutive installations will have the ability to configure all matter, atom by atom, creating conceptually a Utopia of superabundance and vicariously an architecture of cheap and infinitely malleable material. Nanotechnology is a 'bottom up' science, that is starting from the interaction of component parts of a system and generating a complexity from it. 'Top–down' technologies are also pushing the barriers of miniaturization; small machines such as colon crawlers and arterial plaque scrapers are already in prototype form.)

- the increasing use of network solutions (a key driver in replacing products by services delivered over the network)

- the tendency for hardware to be replaced in importance by software

- the increasing emphasis on usability and consequently HCI developments such as GUIs over CLIs and the increasing prominence of computer-related occupation, health, and corporate security concerns (see HCI and User Interfaces in the Glossary for explanation of abbreviations).

Computer systems

Computer systems can be represented by the systems definition as comprising parts — input, system and output. The graphical representation of a system is shown in Fig.1.[3,4] (see also BIOS (basic input/output system) definition in the glossary). The computer system represented in Fig.2 is defined by hardware (the physical electronic components) which require software (or programs) and data media (structured information) for meaningful operation.

To see bodies as parts, particles or bits is not new. For example a Dutch philosopher, Spinoza, writing in the 17th century defined a body as two simultaneous conditions

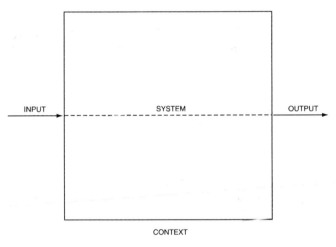

INPUT   SYSTEM   OUTPUT

CONTEXT

Fig. 1. A system as used by Frey, from Reed[4]

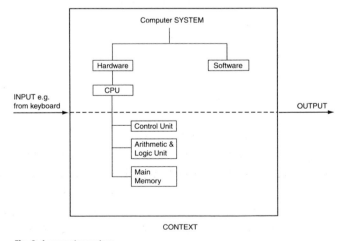

Computer SYSTEM

Hardware   Software

CPU

INPUT e.g. from keyboard   OUTPUT

Control Unit

Arithmetic & Logic Unit

Main Memory

CONTEXT

Fig. 2. A computer system

The body is composed of an infinite number of particles. It is created by the relationship between particles travelling at different velocities.

The body is defined and affected by its capacity to respond to its context.

A brief history of the electronic computer

A brief history of the electronic computer is relevant when considering all these developments, both hardware and software, and subdivision according to several generations is often used. Table 1 charts the first four generations of computer technology.[5]

Computational machines/mechanical devices such as the abacus existed for centuries and as early as the 1800s Charles Babbage (1792–1871), an English professor of mathematics, developed an idea for an analytical engine that could perform any kind of computation automatically. In France, Joseph Jacquard (1752–1832) used punched cards to control the extraordinarily intricate patterns which his loom was capable of weaving, thus presenting a highly versatile input device.

In 1889 Dr. Herman Hollerith (1860–1929), a US statistician and engineer working for the US Census Office, devised electrical tabulating equipment that was used for the 1890 census. Hollerith's system led to many important developments, including the first use of electricity successfully as an integral part of the automation process. His firm, together with the Computing Tabulating Recording Company, later became IBM. His method represented pieces of information by a series of punched holes in cards—now referred to as 'Hollerith' or IBM cards. The hole in the card completed an electrical circuit when it was passed under a wire, and the location of the hole under different wires indicated information.

However, the true basis of advanced Information Technology appeared in the second half of the 19th century with the emergence of the electronic computer as valve-based first generation machines (Fig. 3).[6] These glass 'thermionic' valves which made up the CPUs (Central Processing Units) of the earliest computers filled a room, with

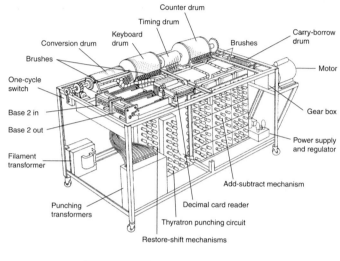

Fig. 3. The first electronic computer (courtesy of The University of Michigan Press)

Table 1. The first four generations of computers

Generation	First	Second	Third	Fourth
Year	1945–55	1956–63	1964–81	1982–
Computer	Vacuum tube	Transistor	Integrated circuit	VLSI[§]
Hardware	Magnetic drum	Magnetic core	Semi-conductor	Optical
Size of system (minimal)	Very large	Mainframe	Mini	Micro
Processing speed	10 KIPS[*]	200 KIPS	5 MIPS[‡]	200 MIPS
Memory size	**2 Kbytes[†]**	**32 Kbytes**	**2 Mbytes**	**250 Mbytes**

[*]KIPS = 1000 instructions per second

[‡]MIPS= million instructions per second

[†]Kbytes = 1024 x 8 bits

[§]VLSI = very large scale integration

Fig. 4(a). Overall view of the ENIAC (courtesy of The University of Michigan Press)[9]

Fig. 4(b) (right). Layout of the ENIAC (courtesy of Reed Elsevier)[10]

produced a lot of heat, developing 400°F temperatures inside the electronic equipment. Burks says that

Not surprisingly, some of the older and more experienced engineers at the Moore school were sceptical about ENIAC. Eighteen thousand vacuum tubes? No one had ever operated a system of more than 100 tubes or so. At any moment at least one tube will be inoperative, and that may spoil the answer! But the electromechanical I/O of ENIAC and EDVAC gave at least as much trouble as the electronics. And so it has remained to this day . . .[8]

each valve as a switch which could be turned OFF by applying a strong negative electrical charge to it, or ON by applying a weak electrical charge. If it was ON, a current could flow through it and if it was OFF no current could flow. Between 1937–1942 the ABC (Atanasoff–Berry Computer) was invented by Professor Dr. John V. Atanasoff at Iowa State University. This entailed modifying the standard algorithm for solving simultaneous equations so that only additions and subtractions were needed, adopting binary arithmetic, and inventing a storage system based on refreshing capacitors. The partly analogue ENIAC (Electronic Numerical Integrator And Computer) took two-and-a-half years to develop, and was completed in 1946 by Dr. John Mauchly and Presper Eckert.

While the binary ABC proved electronic computers could solve intricate mathematical equations (to ten decimal places) the ENIAC showed that they were fast (5000 additions per minute).[7] The ENIAC weighed 30 tonnes, covered 15 000 square feet, was two storeys high (Fig. 4) and was programmed by 6000 simple on/off switches which consumed a lot of electricity and

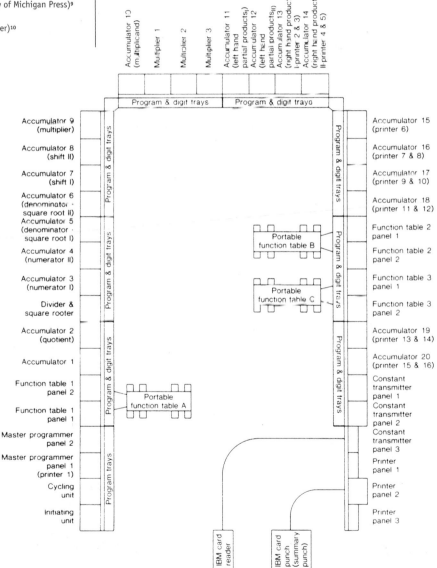

One often-quoted feat was to find the fifth power of a five figure number in half a second, but in doing so its 18 000 valves used 200kW of power. Internal memory by means of magnetic drums meant slow memory access times. This valve-based, unwieldy, machine was largely dedicated to atomic, meteorological and ballistics calculations and tables for the US army. Burks again points out that

The 18 000-vacuum-tube ENIAC proved that electronic technology could be used to make fast, powerful, reliable, general-purpose computers. Its speed of computation was three orders of magnitude greater than the speed obtainable with electromechanical technology, and it solved problems hitherto beyond the reach of man. The electronic design of ENIAC was optimal, led naturally to the stored program computer . . .[11]

The ENIAC provided the western armies of World War II with the basis for a computer which was known as 'Predictor'. The Predictor worked in 3D to calculate when to fire and how long it would take a fired shell or missile aimed at a high-flying aircraft to intercept it. It took into account variables, including the fact that a missile would take 30 seconds to gain the right altitude to intercept the aircraft, the age of the firing equipment, the atmospheric pressures and wind speed both on the ground and where the aircraft was flying. By 1951 this entire process was automated and had a plotter to chart what was happening.

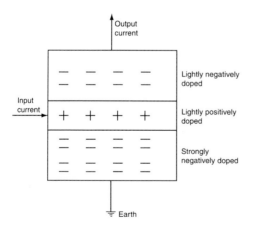

A **transistor** consists of three layers doped silicon, i.e. silicon to which a small amount of another substance has been added. Doped silicon is a **semi-conductor**. This means that under certain circumstances it will conduct electricity otherwise it will not — thus allowing the transistor to act like a switch.

Fig. 5. A transistor

Two computers, both called 'Mark 1' were then developed using the newly invented transistor (Fig. 5)[12] to overcome the major shortcoming of the ENIAC (i.e. the use of 18 000 electronic tubes/valves) and the Predictor (10 000 electronic tubes/valves). Internal memory was by means of core storage. By 1944, while under an IBM sponsorship, a Harvard Professor, Howard Aiken, developed one Mark 1 as a number of electronic calculators joined up in series. Tom Kilburn and Fred Williams developed the other Mark I at Manchester University by 1948.[13] The big advance of this machine was its capacity to store programs, thus benefiting from Kilburn's work on the problem of computer storage.[14, 15] The EDSAC (Electronic Delay Storage Automatic Calculator) was built at Cambridge University in 1949, and represents the world's first stored-program computer. John Von Neumann may be credited with the invention of the computer software now known as operating systems after deducing that the machines, rather than relying on banks of on/off switches being set, could actually store their own programs to read data, process it and produce the desired results.

By 1960 Kilburn had developed the prototype of the giant Atlas computer, with the first commercial machine delivered in 1964 by Ferranti. The UNIVAC 1 (Universal Automatic Computer) for Remington-Rand by Mauchly and Eckert was installed in 1951 in the Bureau of the Census. Sixteen UNIVACs were eventually built and three were used to process the 1951 US Census. By then commercial firms — Burroughs, Honeywell, RCA (Radio Corporation of America) and IBM (International Business Machines) later to be joined by CDC (Control Data Corporation), NCR (National Cash Register) and GE (General Electric) — were also interested. IBM introduced their first commercial computer, the IBM 701, which sold over 1000 — many more than the 50 they had hoped to sell. Figure 6 shows the technologies, machines and ideas, that led, via Atanasoff's computer and the ENIAC to the first stored-program.[16] A comparison of architecture, performance and physical characteristics of the Eckert-Mauchly computers is shown in Table 2.[17]

In the UK in 1953, Lyons, the giant food corporation, was the first British company to invest in computers for commercial data processing, building LEO (Lyons Electronic Office) to keep track of the payroll in its many houses.[18]

The second generation of computers exploited the characteristics of the transistor, whose development was initiated by Bell Laboratories in the USA in 1947. The company had sponsored semi-

Table 2. A comparison of architecture, performance and physical characteristics of the Eckert-Mauchly computers

	ENIAC	EDVAC	BINAC	UNIVAC
Architecture				
Programming	Manual wire panels	Stored program	Stored program	Stored program
Data transmission	Parallel	Serial	Serial	Serial
Number representation	Decimal	Binary	Binary	Decimal
Word length	10 digits	44 bits	31 bits	11 digits + sign
Other data types	—	—	—	12 characters/word
Instruction length	2 digits	44 bits	14 bits	6 characters
Instruction format	1-address	4-address	1-address	1-address
Instruction set size*	97 (100)	12 (16)	25 (32)	45 (63)
Accumulators/ programmable registers	20	4	2	4
Main memory size		1024 words	512 words	1000 words
Main memory type	—	Delay line	Delay line	Delay line
Secondary memory	Function tables	Magnetic drum	Magnetic tape	Magnetic tape
Other I/O devices	Card reader & punch	Cards, paper tape	Typewriter	Typewriter, cards, printer
Error detection	—	Redundant CPUs	Redundant CPUs	Redundancy, parity
Performance				
Clock rate	60–125 KHz	1 MHz	4 MHz[‡]	2.25 MHz
Add time	0.2 ms	0.864 ms[†]	0.285 ms[†]	0.525 ms[†]
Multiply time	2.8 ms	2.9 ms[†]	0.654 ms[†]	2.15 ms[†]
Divide time	24.0 ms	2.9 ms[†]	0.633 ms[†]	3.89 ms[†]
Physical characteristics, approximate measurements				
Vacuum tube count	18 000	3600	1400	5400
Diode count	7200	12 000	n/a	18 000
Power consumption	174 kW	50 kW	13 kW	81 kW
Floor space of computer only	1800 sq. ft	490 sq. ft	n/a	352 sq. ft

n/a: Data not available

* Number of instructions used (number encoded)

[†] Includes memory access time for instructions and operands

[‡] Later reduced to 2.5 MHz (interview with Eckert by Stern, January 23, 1980)

Sources: C. G. Bell and A. Newell, *Computer Structures: Readings and Examples* (New York, 1971); E. C. Berkeley and Lawrence Wainwright, *Computers: Their Operation and Applications* (New York, 1956); Eckert-Mauchly Computer Corporation, 'Engineering Report on the BINAC' (Philadelphia, 1949); Simon Lavington, *Early British Computers* (Bedford, Mass., 1980), p. 125; Martin H. Weik, *A Survey of Domestic Electronic Digital Computing Systems*, BRL Report 971 (Aberdeen, Md., 1955), and *A Third Survey of Domestic Electronic Digital Computing Systems* (Aberdeen, Md., 1961).

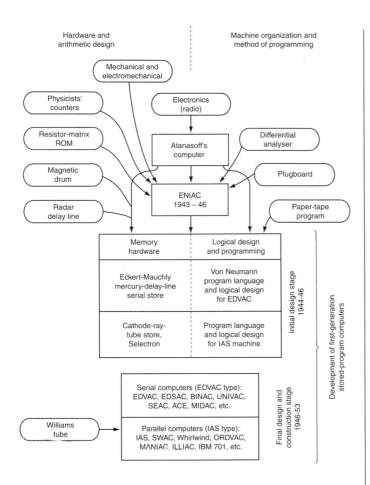

Fig. 6. The technologies, machines and ideas that led, via Atanasoff's computer and the ENIAC, to the first stored-program computers (courtesy of The University of Michigan Press)

conductor research since 1936. As switching devices, transistors were smaller, cheaper, faster and more and more reliable, with fewer and fewer breakdowns than valves.

In 1957 William Shockley was awarded a Nobel Prize for his work on the transistor and subsequently set up the Shockley Transistor Corporation. After disagreements with Shockley, eight members of his staff left and set up the Fairchild Camera and Instrument Company, from which grew the Silicon Valley. By then, and until 1964, the transistor influenced second generation computers and the market was dominated by firms such as IBM and its main competitors; Burroughs, UNIVAC, NCR, CDC and Honeywell, referred to as the 'Bunch' — all those who could exploit semi-conductors. This period saw the emergence of several new computer languages to facilitate the writing of computer applications, notably FORTRAN, COBOL, LISP, RPG and PL/1.

With the launch of the IBM 360 series on April 7th, 1964, IBM heralded the third generation computers characterized by integrated circuits (ICs), thus making the second generation machines obsolete. Lithographic printing methods allowed arrays of minute transistors to be etched on to silicon wafers which were then encapsulated in plastic chips (ICs). The new machines were much faster, thereby helping with the introduction of multi-programming allowing computers to run more than one program simultaneously. These machines offered the prospect of upgrading without forcing too much redesign of their information systems, and their integrated circuits gave rise to minicomputers (ideal for dedicated use in scientific and office work) e.g. DEC (Digital Equipment Corporation) and its PDP series. Developments in IC continued — LSI (large scale integration) of components led to VLSI (very large scale integration). In LSIs there are up to 20 000 components contained on a semi-conductor chip whereas in the VLSIs (90% of which are CAD designed) a single chip can host up to 100 000 components.

The fourth generation computers are probably characterized partly by the introduction of large scale integration which occurred during 1971, but more significantly and specifically by the intro-duction of the microcomputer and its microprocessor of March 1971. Various software innovations were made, including systems network architecture and fourth generation programming languages (referred to as 4GLS for short). Dan Bricklin and Robert Frankston formed Software Arts, and in 1978 launched the world's first spreadsheet written for the Apple Computer: VisiCalc — an electronic balance sheet which became popular with accountants and therefore helped bring computers to the attention of the business community.

In the same year, the microcomputer Apple 11 was launched by Steve Jobs, former Atari employee, and by Stephen Wozniak, former Hewlett-Packard employee (founders of Apple Computer Corporation) as the first real computer enthusiasts could own, and in 1980 the IBM PC based on the Intel 8086 chip was developed.

In the UK, Radio Shack (Tandy) launched TRS80 in 1978 and in the early 1980s Sir Clive Sinclair launched the tiny ZX80 and ZX81 machines which offered advantages because they could be plugged directly into domestic TVs (negating the need for a monitor). They also included the powerful BASIC language which enabled users to write their own programs, and in addition allowed data and programs to be stored and retrieved via domestic cassette recorders.

The shrewd move and launch of the IBM PC in the US in 1981 and in the UK in 1983, targeted at business users and based on reliability rather than price-competition, and 'open architecture', has been significant in allowing IBM to dominate the computer world. However, IBM made a huge mistake and allowed Microsoft to keep the rights to the operating system, which was subsequently licensed to other hardware manufacturers such as Compaq and Dell under the name MS-DOS. That allowed such manufacturers to advertise themselves as IBM-compatible, and made them much more attractive to consumers. Apple, meanwhile, made an equal mistake in judgment by restricting its operating system, which was generally acknowledged to be far superior, to its own Macintosh computers.

A period of leapfrogging then followed in which machines became faster, with more complex, better graphics and more capabilities. In 1984 the Apple Macintosh (an unusual machine in not having an operating system as such), based on the powerful processor the 68 000 chip, was launched. This was followed by the IBM PC AT, based on the Intel 80286 chip, and the IBM PS/2 (short for Personal System/2) in 1987.

The Apple Macintosh was based upon Rank Xerox's original painstaking research work on the computer working environment of the 1970s involving

- the use of the mouse (to produce corresponding movements of the cursor on the screen rather than typing at the keyboard)
- the use of graphical display (more user-friendly and less difficult to use than text-based display)
- the use of icons (pictorial representations, e.g. wastepaper basket to represent delete/get rid of files, a filing drawer to represent all the files on a disk drive).

It benefited from the appearance of some applications packages which made full use of its graphical environment and led to desktop publishing. Another useful innovation was the desktop laser printer which had made its appearance by the mid 1980s. The IBM PC AT and PS/2 were both based on the same Intel chips and incorporated microchannel architecture and superior VGA screen display. They also offered a Macintosh-like computing environment known as 'presentation manager'.

Fifth generation developments continue, but are often associated with developments relating to artificial intelligence, expert systems (application to strictly defined cognitive activities), natural language computers (which respond to language as written or spoken), RISC (Reduced Institution Set Computers) and transputers (developments in which entire computer systems rather than just components are etched upon single silicon chips). Other developments relate to improved human-computer-interaction (HCI) and therefore usability of computer systems, e.g. user interfaces such as CLI or GUI. IT is therefore divided into four parts: data access, inference machine, user-friendliness and intelligent programming.

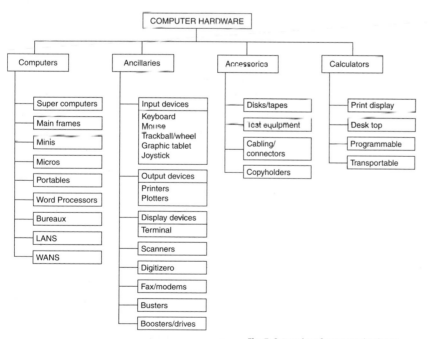

Fig. 7. Categories of computer hardware

Computer hardware components

Computer hardware refers to the physical components of the system. These are depicted in Fig. 7. Classifications of computer systems are distinguished according to the degree of processing power personal computers (PCs or microcomputers), minicomputers, mainframe computers and super-computers. PCs offer a limited degree of processing power to a single user for a relatively low cost. Minicomputers, originally developed on the principles on which mainframes were designed, such as the IBM AS/400 series (DEC VAX), offer more data processing power and provide multiple-user access via dumb terminals (keyboard and display screens) to central storage facilities (floppy disk drive).

Mainframe computers used in organizations such as banks offer even more data processing power and provide multiple-user access to central storage facilities. The term derived from the huge number of racks which were necessary to house the extensive volume of equipment which comprised early computers. The 'main' frame(s) supported the central processing unit (CPU). Other racks (or frames) supported the peripheral equipment. Now the term is still used for large computers — particularly in order to distinguish between micros, minis and mainframes. Batch (as opposed to interactive or real-time) processing is employed in order to free computer capacity in the day for activities requiring instantaneous data processing.

The very fast, very powerful, super-computers with a very large main store such as Cray-2 (after the developer Seymour Cray, a former Control Data Corporation employee) used in military research, in weather forecasting, in analysis of fluids in test tunnels and in census data analysis which involve complex problems requiring huge amounts of processing, offer even more data processing power. Literally billions of calculations and data manipulations (in which information only has to flow down the shortest possible lengths of internal wiring) are handled so quickly that liquid nitrogen is circulated around the physical components of many super-computers to stop them overheating. Although the definition of super-computers is vague and ever-changing (especially since the advent of cheap and easy parallel processing), all the same, the names of Cray and Cyber still remain supreme among leading super-computer manufacturers such as Fujitsu, NCUBE, Mas Par Computer Corporation, Intel, Silicon Graphics and Network System Corporation. Thus the Cray's super-computer of the early 1990s, based on gallium arsenide rather than silicon chips, has a clock speed of 2 GHz, tens of times faster than others.

Main areas of computer hardware specification are

- CPU type (microprocessor — the silicon chip that manipulates data in the computer)
- CPU speed — that is the processor speed at which a computer runs according to (a) the speed of its internal clock, measured in millions of cycles per second (MHz — megahertz) and (b) the average number of clock cycles it requires to execute an instruction (a 386 chip has a clock speed of 20 or 25 MHz and will require about 4.5 clock cycles to perform an instruction giving a processing speed of 20/4.5 = 5 MIPs (million instructions per second)
- RAM (Random Access Memory) capacity — that is the temporary store for holding programs and data loaded from disk or typed at the keyboard or input from some other device
- graphics format (e.g. VGA, etc.) providing the graphics facilities.

Systems software

At this point it is convenient to define/note the key elements in general software types: operating systems (OS); environments; utilities; memory-resident software and applications.

Languages are classified according to whether they are low-level or high-level as shown in Fig. 8. In the early days of computers in the 1950s, programmers used special language consisting of very long sequences or strings of binary 0s and 1s known as machine code (the code or language the machine understands and created by computer manufacturers to fulfil the needs of their particular machine's circuitry for interpretation as instructions). However, this (now referred to as the first generation language) was difficult to read, test and amend, also time consuming and laborious to write correctly even for simple tasks. Therefore assembly code or second generation language (consisting of a set of mnemonic codes, symbolic of the binary instructions which they represent) was developed to translate simple instructions directly into strings of 0s and 1s of machine code and generally to speed up the task of programming.

Usually the assembler translates instructions from the assembly language source program into instructions in the machine language object program on a one-to-one ratio. Commands such as LDA (load accumulator), ADC (add with carry) and CLC (clear carry) were a feature of the assembly code.

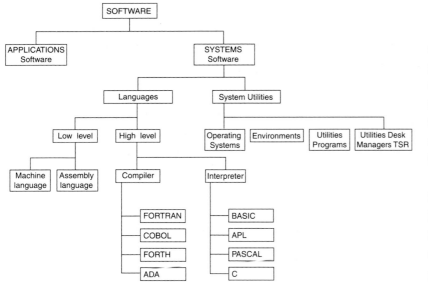

Fig. 8. Levels of computer software

Both machine code and assemblers or low-level languages are machine-dependent (they address the actual circuitry inside the CPU of a computer — varying from computer to computer). Therefore an intimate knowledge of the way the computer works is required, including every aspect of the storage and manipulation within the computer of the instructions and data.

To reduce further the amount of work required to program a computer, high-level procedural languages — HLLs (or the third generation compiled languages, for example COBOL, FORTRAN, PL/1, BASIC, Pascal and C) were developed. The program for the translation of the HLL source program into the machine language object program is referred to as a compiler — although sometimes an interpreter is used. A compiler is more elaborate in its operation than an assembler, although it performs a comparable function.

The HLLs allowed the programmer to work at a much higher level, isolating him/her from the operation of the computer and resulting in programs which are easy to write, read, test, correct and expand, and which also allow faster programming. Knowledge of computer circuitry is not needed to write programs. Instead, programmers could concentrate on the overall structure of the program so that the computer could translate formulae and other features into the more basic instructions used by assemblers, and finally the resulting code could be assembled into the low-level machine code. One of the main problems with procedural languages is that each instruction forms part of a sequence, so that the programmer must

not only write correctly but also anticipate every possible set of conditions. If error conditions or a set of conditions which the programmer has not anticipated occur and there are no instructions on how to deal with these, then either the computer will stop running or will do something at random, almost invariably something unwanted.

Because people do not work well with procedural languages, fourth generation declarative languages, known as 4GLs, such as SQL (Structured Query Language used in databases) were developed to overcome these shortcomings. They compel the programmer to define the procedure, step by step, to achieve a result. 4GLs are non-procedural (they tell the computer what to do rather than how to do it). SQL (available for Paradox, DataEase, Oracle, Ingress, Microsoft Excel) allows users to communicate with servers. In displaying data it is only necessary to describe how the data are to be displayed, rather than the methods used to display them. There is no need to specify how to carry out tasks but simply what tasks need to be done. It is not necessary to learn a complicated non-English syntax in order to use the language.

The most common fourth generation languages are closely linked with the development of the visual display unit (VDU). Until the VDU was invented, all interfacing to computers was done via teletype terminals. These printed a line at a time and did not allow the user to move quickly and easily to different parts of the screen and provided limited response. Declarative languages, although quite user-friendly and providing interactive dialogue with users, with a number of aids, such as screen formats and menus, to ensure that users do not need to learn complex languages, do not work well with computers. They must first be translated into procedural languages, thereby increasing the possibility of incorrect translation.

The following list of the main languages is in order of the year of availability.

- *1954 FORTRAN* (FORmula TRANslator) features mathematics and is based on Turing Machine Theory. The first high-level language, produced by John Backus and others, FORTRAN's only data structure is the array (or matrix), and the data types include double-precision and complex numbers.

- *1959 COBOL* (COmmon Business Oriented Language) is commonly used for the development of business applications with large volumes of data and summary form of outputs. Also based on Turing Machine Theory it features data processing and consists of four divisions: (*a*) identification (information for users); (*b*) environment (hardware); (*c*) data (files to be used), and (*d*) procedure (operations to be performed).

- *1960 LISP* (LISt Processor) (most commonly used for symbolic computation) is second to FORTRAN as the oldest programming language in general use. It is based on lambda calculus and recursive function theory. It requires a minimum of 40 megabytes of memory. The LISP environment consists of a language, a programming environment and a set of 12 000 functions.

- *1964 RPG* (Report Program Generator) incorporated in many database management programs, it helps the non-expert user define the reports required so that the system can produce the correct report programs.

- *1964 PL/1* (Program Language Number 1) is a high-level program language of value in both business and science contexts. It was devised by IBM with the aim of making it so flexible that there would be no further need for other program languages, but this did not prove to be the case. PL/1 was designed to combine the best parts of FORTRAN, COBAL and ALGOL, i.e. to overcome the inadequate handling capabilities of FORTRAN and the inadequate scientific calculation capabilities of COBOL.

- *1965 BASIC* (Beginners' All-purpose Symbolic Instruction Code) is a high-level programming language suitable for on-line program development and popularly used to introduce beginners to programming techniques. Originally designed in 1964 by John Kemeny and Thomas Kurtz (Dartmouth College, US) for non-interactive use on mainframes.

- *1968 Pascal,* originated by Niklaus Wirth and colleagues at the Institute of Informatics, Zurich, is suitable for teaching structural programming techniques. It was named after the French mathematician and philosopher, Blaise Pascal (1623–1662) and features portability.

- *APL* (A Programming Language) is a powerful programming language which can handle great power arrays of several dimensions. It features mathematics and data processing.

- *FORTH,* named as a fourth generation language, was published in 1968 for telescope control by Charles Moore, a US astronomer. It consists of separate words (instructions), each defining an action. The system compiles each word straight after entry, so flags errors at once. It stores all data in a stack making it fast and compact, and suitable to control. It features process control.

- *1970 PROLOG* (PROgramming in LOGic) is a declarative non-algorithmic language. Alain Colmerauer and others first produced it in France in 1972 although the main progress since has been in the hands of Robert Kowalski in the UK and the Japanese ICOT centre (which uses it for the national fifth generation development programme). Its specialties are rapid application prototyping system-level software and a backward chaining expert system. It is relevant for expert systems allowing users to add facts and rules to the knowledge base of facts and rules and cannot be used to write programs. An expert system consists of three components: (*a*) a knowledge base of facts and rules gleaned from human experts about a particular domain (field); (*b*) a knowledge manager software which controls the knowledge base and (*c*) situation model containing data on the current situation or case. PROLOG is a higher level language than LISP, the other symbolic processing language. It uses parallel processing (examines all the rules at the same time).

- *1972 C* (developed by Bell Laboratories for use in writing systems' software) is a compiled language created to develop the UNIX operating system, and spans the gap between high and low-level languages.

- *1980 ADA* (US Defense Department Mandated Program Language) features robotics, payroll and radar programming. ADA has Pascal roots and its main strength is in the structured coding of real-time systems and is best known for its use in nuclear power station control and US weapons systems, but is of fast growing value to commerce.

- *1991* The first *World Wide Web* (www) browser was developed by researchers at CERN particle physics laboratory in Geneva, Switzerland. Instead of displaying simple text, the browser retrieved hypertext documents from the host computer and displayed these documents in glorious colour, in a variety of fonts and with embedded graphics.

- *1995 Java* was developed by Sun Microsystems as a programming language that can be read and translated to run on any platform. Java allows the creation of interactive multimedia applications delivered over the network. Internet applications can use sound and 3D graphics yet be platform-independent, running on UNIX, Windows 95, Windows NT and Mac. The alpha version of the Java language and the hot Java browser were released in the spring of 1995 and by the summer of the same year Netscape had licensed Java Technology for use in its market-leading web browser. Java 1.0 was released in 1996.

FORTRAN, COBOL, ALGOL, Pascal and ADA are based on Turing Machine Theory developed by Alan Turing (1912–54), a noted English mathematician who designed one of the first electronic computers to be built. In 1937 he wrote about a hypothetical machine that could use a system of binary codes to perform any algorithmic operation. This was the conception of the universal computer and was called the 'Turing machine'. In essence he devised a test (appropriately called the 'Turing test') that determines whether a machine can think. The term 'recursive' describes a technique in which a circular process is used to perform an iterative process. In other words the value of a function is derived from a more elementary value of the same function.

Operating systems refer to the computer programs which will run the computer by organizing the memory, send instructions to any peripheral devices and protect the memory from errors in the user's program. Examples are MULTICS, IBM's OS/360, TENNEX; for small computers, CP/M, MS/DOS (licensed by Microsoft on the original IBM PC in 1982) and OS/2 from Microsoft (introduced in 1987 by Microsoft and IBM as the successor to DOS — this was abandoned by Microsoft in 1990 and Windows NT was originally introduced as OS/2 3.0 in November 1987 and the first version (1.0) was released in 1988 while IBM continued development of OS/2 with the third major release of 'Warp' in 1994), DOS (Disk Operating System) and UNIX from AT&T. Carter[19] groups the tasks of operating systems into five headings

(a) basic input/output operations, e.g. control of the screen display using BIOS (Basic Input/Output System)

(b) disk operations to do with storing programs and data on disk using DOS (Disk Operating System)

(c) network operations

(d) multi-tasking operations which enable handling of several tasks at the same time (e.g. background printing)

(e) multi-user operations which permit a number of users of the computer and its software at the same time.

Table 3 shows CPUs, their manufacturers and operating systems.

Environments are programs which act as a layer between the operating system and themselves. Examples are IBM's Environmental Manager, Apple's WIMP, Digital Research's GEM and Microsoft's Windows (a 'shell' run 'over' MS-DOS). Such environments tend to use pictorial representations of commands, e.g. the GEM picture of a rubbish bin to

Table 3. CPUs, manufacturers and operating systems

	CPU	Manufacturer	Operating System
1980s CPUs	8088, 80286, 80386	Intel	DOS, Windows, UNIX
	68000,68010,68020	Motorola	MacOS, UNIX
1990s CPUs	80486/Pentium/P6	Intel	DOS, Windows, UNIX, Windows NT
	68030, 68040	Motorola	MacOS, UNIX
	SPARC	Sun Microsystems	UNIX
	Alpha	DEC	VMS, UNIX, Windows NT
	PowerPC	IBM/Motorola	MacOS, UNIX, Windows NT
	PA–RISC	Hewlett-Packard	UNIX, Windows NT
	R3000, R4000	Silicon Graphics	UNIX, Windows NT

indicate 'erasefile'. Microsoft has an iron grip on the corporate desktop through its MS-DOS, Windows 3X, Windows 95, 98 or 2000 and Windows NT workstation operating systems. Microsoft's technological prowess is only matched by Intel's dominance of the global market for microprocessors, the brains for PCs. Licensing the technology to IBM for what later became the MS-DOS operating system is a classic example of a match strike. MS-DOS subsequently became the world's most popular computer operating system and Microsoft set off on its path to becoming the world's largest independent software company. Harvard drop-out William Gates III and his school friend Paul Allen, both from the Seattle area, founded Microsoft in 1975 to sell a version of BASIC programming language. Microsoft's revenue reached a billion dollars in 1990, and it is now a worldwide business with approximately 20 000 employees, operations in some 50 countries and revenues of US $8.67 billion in the fiscal year 1996. Systems software still accounts for approximately one-third of the company's revenue and applications software accounts provide the other two-thirds. Gates often stated that part of Microsoft's mission statement was to put a personal computer on every desktop — a mission that has been amended recently to include putting a server under every business desktop. This is not mission impossible.

Utilities refer to programs that transfer files of data between computers or small programs that allow one use of the computer in more powerful ways. They extend the power of the system by carrying out tasks beyond the capabilities of the operating system and carrying out operating system tasks in a more efficient and easier manner. They are run from the operating system itself and are therefore referred to as front-ends. Common examples include PowerMenu, PC Tools, Qdos and Xtree to DOS. The term utilities has also been used as a synonym for 'desktop management'.

A Lotus organizer is a program that contains all or some of the following: a calculator; a calendar and an alarm clock, a 'notepad', a telephone and address list, a telephone dialer, an appointments diary, an elementary text editor, filing system and index card file, plus a list of things to do.

Memory-resident software (TSR — Terminate Stay Resident) refer to utilities desk managers which can be put into memory and left there in such a way that they only come into use when needed. After use they stay quietly in the background. Disadvantages of TSRs are that they can end up in conflict with each other and can lock up a complete system. Applications would include such areas as accounting/book-keeping, marketing, spreadsheets, database management, personnel, graphics, financial modelling, production, hypertext, word processing, statistics, communications and industry specials.

Applications software

Applications software refers to programs for specific tasks (for example word processing, office automation, business graphics and presentation, painting and animation, project planning, payrolls, duct sizing, computer-aided design and manufacture, and so on), all of which are distinguished from general software mentioned above. Such software is normally supplied as a package on floppy disks with explanatory manuals and a tutorial, either on disk or in booklet form or both. Most application packages may be command driven (using DOS) or menu-driven (using Windows and GEM) or as a combination of both.

Whereas organizations would once have bought all their computer equipment and software from one supplier, they are now buying from several. Typically, everyone bought the box from IBM and software from a single supplier. Increasingly, the trend is to multi-vendor, requiring self-awareness and self-reliance from the end-user rather than the provision of wall-to-wall service and back-up as in the past.

International Data Corporation (1983) showed that office workers, among others, spent most of their time doing written activities (managers 28%, others 15%, office workers 17%). The range of the activities (e.g. writing, taking dictation, reading, proofreading, searching, filing, copying, distribution, operating equipment, etc.) indicates the differing areas in which Information Technology applications might be used. In Table 4 Mitchell[20] lists, according to different stages of the design process and based upon the results of surveys (Hoskins 1973, Lee 1974),[21-23] the range of functions for which useful programs have been developed.

This list still applies today although it could be expanded to include facilities management and post-project activities, for example Cornick mentions element maintenance programmes,[24] while the computer applications in *Computer applications in architecture*[25] cover

- economic feasibility studies for building development

- space planning

- building appraisal

- environmental design and building services

- specifications

- data handling and manipulation

- graphics

- the ARK 2 system — the first major interactive design system

- the OXSYS system — centred on the Oxford Regional Health Authority's Oxford Method of Construction.

More recently, Mitchell and McCullough in *Digital Design Media*[26] organize their discussion of software not by category or task, but by the dimension of the media manipulated by the software

- 1D media including words, text and sounds

- 2D media including images, drafted lines, polygons, plans and maps

- 3D media including lines in space, surfaces, renderings and assemblies of solids

- Multi-dimensional media including motion models, animation and hypermedia.

The range of applications software for architects and engineers form the subject matter of this book. The emphasis is on the

Table 4. Range of functions for which useful programs have been developed

Briefing phase

	Feasibility study	Economic feasibility analysis, Housing type mix analysis
	Programming	Problem structuring, Activity data analysis, Space needs provision, Accommodation schedule production, Circulation analysis, Cluster and bubble diagram generation

Sketch design phase

	Site planning	Site mapping, Slope analysis, Drainage analysis, Cut and fill analysis, View analysis, Accessibility analysis, Overlay mapping analyses, Site plan synthesis
	Schematic design synthesis	Floor plan layout, 3D spatial synthesis
	Performance & cost analysis	Checking for compliance with the brief, Circulation analysis, Preliminary structural computations, Heat gain and heat loss computations, Insulation and shadow pattern analyses, Natural and artificial lighting computations, Sound transmission and reverberation time computations, Preliminary cost estimation
	Presentation	Plotting sketch plans, elevations, sections, perspectives

Production documents phase

	Detail design	Building products data retrieval, Automated detailing, Structural member selection and sizing, Mechanical and electrical system detail design, Duct, pipe and electrical network layout
	Costing	Generation and pricing of bills of quantities, Cost analyses
	Production	Generation of schedules and specifications, Plotting of working drawings

Construction supervision phase

	Network analysis, Precedence diagramming, Project cost control

Management functions

	Job costing, Time and payroll functions, Invoicing

usefulness of software rather than a preoccupation with the vexed question of which hardware or software to use.

References

1. Ostman C. Nanotechnology: the next revolution. *21st Century On-line Magazine*, http://www.21net.com

2. Drexler K. E. *Engines of Creation*. Oxford Press, 1992.

3. Frey H. *Design strategies*. Doctoral thesis, University of Strathclyde, Glasgow, 1989, A39.

4. Reed P. *Engineering Services*. In Markus T. A. and Morris E. N. (eds), *Buildings, climate and energy*. London, 1980, Ch. 11, 415f.

5. Zorkoczy P. *Information technology, an introduction*. Third Edition, Pitman, 1990, 97.

6. Burks A. R. and Burks A. W. *The first electronic computer*. University of Michigan Press, Ann Arbor, 1988, 52.

7. *ibid.*, 311–344.

8. *ibid.*, 312.

9. *ibid.*, 113.

10. Stern N. *From ENIAC to UNIVAC: an appraisal of the Eckert-Mauchly computers*. Digital Press, Bedford, Massachusetts, 29.

11. Burks A. R. and Burks A. W., *op.cit.*, 335.

12. Carter R. *Information Technology*. Butterworth-Heinemann Ltd, Oxford, 1991, 22.

13. Metropolis N., Howlett J. and Gian-Carlo Rota, (eds). *A history of computing in the Twentieth Century*. Academic Press, New York, 1985, 440.

14. Kilburn T. *A storage system for use with binary computing machines*. Doctoral dissertation, Manchester University, 1947.

15. Stern N., *op.cit.*, 155.

16. Burks A. R. and Burks A. W., *op.cit.*, 281.

17. Stern N., *op.cit.*,133.

18. Martin C. and Powell, P. *Information systems a management perspective*. McGraw-Hill, Maidenhead, 1992, 52–53.

19. Carter R., *op. cit.*, 96.

20. Mitchell W. J. *Computer-aided architectural design*. Van Nostrand Reinhold Co. New York, 1977, 75–77.

21. Hoskins E. M. *Computer-aided building: a study of current trends*. Applied Research of Cambridge, Cambridge, 1973.

22. Hoskins E. M. Computer-aids for system building. *Industrialization Forum*, **4**, No. 5, 1973, 27-42.

23. Lee K. (ed.) *Computer programs in environmental design*. 5 vols. Environmental Design Research Centre, Boston, Massachusetts, 1974.

24. Cornick T. *Computer-integrated building design*. E & FN Spon, London, 1996, 81.

25. Gero J. S. (ed). *Computer applications in architecture*. Applied Science Publishers Ltd, London, 1977, 6–10.

26. Sanders K. *The digital architect — a common-sense guide to using computer technology in design practice*. John Wiley & Sons, New York, 1996, 71.

Text and graphics image manipulation

Introduction

Administration activities concern production, alteration and storage of the written word, for example production of letters, memos, reports and documents in the three main market segments — corporate, professional and personal. This allows different word processors to offer different facilities — some are particularly suitable for writing books and major reports, others for work involving the extensive use of scientific and mathematical symbols, and yet others for general small office work.[1] The use of word processors has revolutionized all aspects of writing. It is by far the most common application of PCs.

Word processing

The main advantage of word processing is the reduction in re-types. The original typing is stored in memory, so that any alterations or errors can be effected and the result printed out automatically at speed. By so doing it enables one to think and edit (change) ideas in a manner that is not possible using a typewriter. Book authors who use word processing would not wish to change to typewriters. However, the preparation of commercial reports in an office which maintains a secretarial pool may be undertaken more efficiently by means of dictation. The main disadvantage of centralized word processing occurs when the central processor stops operating, an event that is positively panic-producing.

The early word processors were what would now be called 'automatic typewriters'. The development of word processing has been from the technical end towards the user end. The machines used for word processing developed from typesetting equipment, computing, data processing and automated typesetting, merging into word processing.

Typography is the output of word processing; it is the words, the alphanumerics, the alphabet, numerals and symbols of our graphic form of word communication. IBM introduced the term 'word processing' with the MTSC (Magnetic Tape Selectric Composer), such as the varityper, developed in the 1930s. In so doing, word processing was immediately linked with typesetting because a composer is a typewriter-like device that produces a type image by direct impression. In the varityper, the original composer, the keys are arranged like those of a typewriter but accommodate extra symbols above the capital letters. The image it produces is like that of a carbon-ribbon typewriter but is actually that of a font of type. With a selection of interchangeable type fonts and the capability to adjust the machine for variables, type can be composed. In the MTSC, the tape is transferred to the record-reading unit at the left of the composer, and played out at 14 characters per second and 150 words per minute to produce typesetting composition. The typeset configuration or format is entered on the console. The word processing machine is both a typewriter that records keystrokes and a system of information management.

The main function of automatic typewriters was the utilization of standard paragraphs in letters, reports and contracts. The advantages of such systems over the previous shorthand-typing approach was that it was not necessary to read the letters, reports or contracts for typing errors, thus speeding up the whole process. An organization knew that the letters were well written because of the use of previously approved standard paragraphs. For instance, in the 1950s Unilever used this type of system in the public relations field to reply to public inquiries concerning visits to their factories. It was only necessary to insert the person's name and the relevant dates. It was also possible to 'mail-merge' (i.e. to send out standard letters to a list of people whose names and addresses were held separately). Mail-merge is thus a process whereby a personalized standard letter (typed in the first document file) may be sent to every person on a mailing list, with each recipient receiving correspondence printed with their own name and address (typed in the second document file).

Today, word processors can do a lot more. Essentially, word processors are text-manipulation programs. They can edit (add, remove, alter or move the position of words, phrases, paragraphs or sections of text at will), automatically number pages, sections of text and paragraphs, and subsequently renumber for revisions. They can include automatic collection and printing of footnotes and textual references, and automatic indexing of key words so that the index will include cross-references as well as page numbers.

In addition, they can provide automatic word wrapping (i.e. the process whereby text can be continuously typed into a word processor without the user having to press the carriage return key at each line end) and they can allow justification (right, left and centre); make margin and top-and-tail adjustments; mix single and double spacing or typefaces/fonts; search the text for a string of characters and then replace them with another (e.g. replacing all references to IT with Information Technology). They allow mistakes to be corrected, routine typing errors to be detected, check spelling automatically from a dictionary and they can prepare a table of contents from the main headings.

These features have meant that today's word processing software has three main components (Fig. 9).

(a)   the *text editor* (the part that enables the user to type or enter text and change or edit the material)

(b)   the *formatter* (the part that allows the user to decide the appearance of the final document or other input)

(c)   the *outputter* (the part that controls the printer or other output device).

Fig. 9. The three main components of word processor software

Examples of word processing packages are:

- WordPerfect® (Corel Corporation)
- Microsoft Word® (Microsoft Corporation)
- Lotus Ami Pro® (Lotus Development Corporation)
- MultiMate® (acquired by Ashton-Tate)
- Protext® (Electronic Transcript Software — ETS)
- WordStar® (Micropro)

WordStar, a DOS program, had been the most successful word processing package from the beginning of the PC in 1981, establishing itself early in the game right up to the mid 1980s. Then it began to lose direction; the company that made it, Micropro, tried to launch a new version called WordStar 2000 that no-one really liked, so it had to go back and revamp the old version of WordStar.

WordPerfect, also originally a DOS program, had been coming up on the inside and just ploughed on through. It was much more powerful and supported most printers that existed (and a few that did not). It became the most used word processing software in the world with over 15 million users. But problems arose when WordPerfect launched its word processor for Windows. It was a disaster, and a situation made worse by comparison with Microsoft's Word for Windows which was in its second release and just about getting into its stride. WordPerfect never really recovered.

Developments in word processors have led to complex features such as sophisticated text editing programs comparable to CAD software.

Desktop publishing (DTP)

While word processing programs are effectively based upon the manipulation of an endless scroll of text that is eventually formatted into pages or output, the more extensive capabilities of desktop publishing programs are page-based, and optimized for text and graphical layout.

Apple, with the launch of the Macintosh and MacDraw software, not only led the way in propounding and developing DTP but were instrumental in advocating and implementing what have now evolved as the acceptable industry standards — Postscript support and WYSIWYG (what you see is what you get) interface. The Macintosh offered an innovative, small, one-box solution, with closed architecture, no add-on boards, no options, providing a rather humble 128K of RAM and 400K disk drives. Key elements of a DTP system are

(a)   an operating system with integral WYSIWYG user interface

(b)   a minimum of three software packages — WP (word processing), graphics and page composition

(c)   the facility to support a high resolution output device (printer, plotter, etc.).

Figure 10 shows the constituent parts of a typical single-user DTP system. DTP is a generic marketing term for systems which can accept keyed input and may scan in graphics to a microcomputer, make-up pages with varying degrees of flexibility and graphical facility, and output the results on a laser printer or other similar high quality output device using industry standard typefonts. Three functions are input, make-up and output.

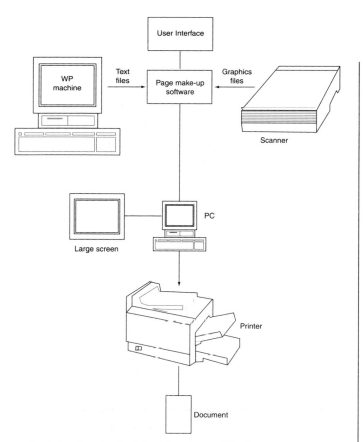

Fig. 10. Constituent parts of a typical single-user DTP system

This means that preparation of text data files in a word processor allows import into DTP programs and optimization for final page layout (columns, headlines, running headers and footers, etc.). Comparison between the traditional process of book production and DTP is shown in Fig. 11.[2] There are similarities between DTP and typesetters of the printing industry in that both employ laser technology to build up characters (Raster images) allowing convergence of technologies. There is a feedback between the two technologies — DTP takes its standards in both resolution of output and quality of typographical precision from the traditional print and publishing industry. A look at historical events notes that the graphic developments centred around typesetting. The linecaster was introduced near the turn of the century and it mechanized the composition of type.

It was the production tool used by everyone in the graphics industry for typesetting. As automation was introduced to this field, the concept of typesetting went through a series of generations. In the first generation, early photosetters were modelled after hot metal machines and were largely mechanical operations. They used film machines instead of metal matrices and adapted the mechanized operation of the linecaster from metal typecasting to film-print typecasting. This improved the speed of the linecaster, as

Application areas for DTP are

- the professional printer and publisher

- the organization requiring a considerable amount of typesetting (often a service provided by external agencies)

- the office user wishing to enhance the presentation of written communications. DTP systems emulate the skills of the professional compositor. These operations are similar to the functions performed by specialists such as the traditional page layout artists in newspaper publishing, characterized by 'cut and paste' to arrange columns of text, headings and graphics images on a page-by-page basis.

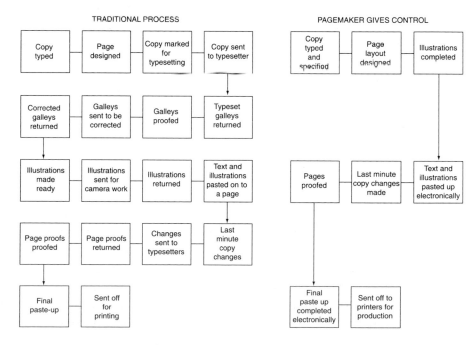

Fig. 11. Comparison between the traditional process of book production and DTP

the casting and handling of slugs made with molten metal was eliminated.

The second generation was characterized by photosetters such as Proton phototypesetters using electromechanical means of exposing typefonts. Use was made of film to produce the type image but also integrated the mechanism with the computer, thereby requiring a number of perforating keyboards to feed one typesetter. In general it was used for applications requiring quality typography. In the third generation, photosetters were using cathode-ray tubes to generate the typographical images (i.e. to generate the image for optical reproduction). This is a fast process which allows use in volume applications, with speed no longer measured by the parameter of typing speed but in lines per minute. In the fourth generation, photosetters are characterized by the use of lasers to expose characters.

DTP programs have a greater range of fonts and typestyles, and provide a more accurate control of line and letter spacing, colour usage, text and graphics rotation than most word processing software. DTP software packages which have been developed include

- Aldus PageMaker® by Adobe Corporation (page-based)
- Ventura Publisher® by Corel Corporation (document-based and for lengthier texts)
- PagePlus® by Serif SPC
- QuarkXpress® previously distributed by Heyden.

A wide variety of desktop published brochures can be produced using, for example, Pagemaker® in the Apple environment or QuarkXpress® in all other platforms.

DTP offers four advantages in areas of form production, graphics/artwork design for logos, presentation material and diagrams

(*a*)   cost benefits

(*b*)   time savings (shortens the proofing cycle)

(*c*)   control (offers ability to edit) of creation and input into layout and printing which allows for author alterations while maintaining control over content and aesthetics between originator, typesetter and printer

(*d*)   image/profile (upgrades company standards and level of presentation).

DTP is therefore useful for producing reports as quality creative documents, especially if other data such as drawings, charts and graphics need to be incorporated. It can be used for creating business stationery; printing a couple of dozen sheets at a time, which allows for easy changes when necessary. Notwithstanding all this, for some architectural/engineering practices who outsource these activities DTP software may be over the top, word processing may be all they need. Increasingly developments, e.g. merging or linking drawings and images in documents in word processing packages such as Microsoft Word, mean that the distinction between DTP and word processing is blurred and irrelevant.

Graphics manipulation

The application of presentation graphics software ranges from the production of overhead and even 35 mm slides to the creation of diagrams for import into word processed or DTP documents. It is widely accepted that graphs convey trend, results and any messages more effectively than a string of figures or numerical data.

Business graphs offer the user savings in labour and in time to produce financial and statistical reports, presentation materials, published accounts and so on. These presentation graphics packages are not suitable for complex design or illustration work because they are intended for use in a variety of ways. Examples include the production of a range of charts (pie, bar, stacked column, line, area and 3D charts, etc.), and graphs (bar-bullets, lists, scatter, xy, etc.). They also enable the addition of titles, arrows, labels (using simple drawing tools which enable the construction of figures from lines, circles, boxes and other basic shapes), and can produce graphs derived from user data.

Examples of the typical software available include

- Harvard Graphics® by Serif SPC
- Freelance® by Lotus Corporation
- CorelDraw® by Corel Corporation.

Presentation graphics packages produce structured or vector-based images, i.e. their drawings are stored in a precise mathematical form of co-ordinates and lines. Once constructed, each drawing element can then be manipulated or resized any number of times without losing any of its detail — a distinct advantage if the image creations are to be exported between different software programs, with their size being tailored to suit the final layout under consideration.

It is important to make the distinction between the following: a vector image, a raster image, vectorization and redlining (Fig. 12). On the one hand a vector image is created by drawing lines from point-to-point so that such lines are mathematically defined and are stored as co-ordinates of their end points. Since they are mathematically precise, the drawing can be zoomed and zoomed again.

The lines are calculated according to how much you see and, in a CAD program, the line will normally be drawn with a single pixel thickness. AutoCAD stores its drawing data in vector form. A pen plotter outputs vectors by using pens to follow the positions of the lines.

A raster image, on the other hand, is created by a pattern of dots which are not connected, as can be seen when the image is enlarged (i.e. as the dots spread out the line gets thicker). Scanners produce raster images since they work by scanning and positioning dots in the image where the scanner sees a line or point. Raster output devices — such as laser, inkjet and thermal plotters — are much faster than their vector equivalents, although quality, since the image is made up of dots, may not be as good.

Vectorization is the act of converting scanned raster images into CAD drawings. This can be done by hand, by tracing or digitizing a hard copy drawing, or by using software to analyse the patterns of dots and produce vectors where it sees a pattern forming a line. This can be a slow task and can give rise to errors where there are a lot of marks on the scanned image. So a first step is to remove marks added to images by a scanner, or unwanted marks on the hard-copy image. Paper warps and stretches, so the image may need to be de-skewed or rubber sheeted where a part of the image can be pulled into place. Text is always a problem with vectorization, since it normally consists of a lot of small lines and one cannot risk an OCR (Optical Character Recognition) program mistaking key numbers in dimensions or words in specification. Such elements often need rekeying. Redlining refers to adding text or components to a drawing or image as a separate layer so that others can obtain this information. These notes are normally in a colour which stands out, hence the term redlining.

Designing and using business/standard forms

Forms (e.g. timesheets, expenses forms, etc.) are an important area in the day-to-day activities of a business and design of these can

benefit from computerization techniques. Special purpose programs are available for this task. These form generators use the interactive power of the VDU.

Specification writing

This section is concerned with *specifications*, an important area in both architecture and engineering and one of the formats describing information within the design process. The other formats are *layout drawings* (plans, elevations, sections, perspectives, etc.), *schedules* of accommodation requirements, quantities, finishes, doors, windows, etc. and *detail drawings*.[3]

In specification writing it is important that there is fast and accurate production of project documents plus sophisticated management of the technical content of specifications. It seems appropriate that this should be considered after looking at word processing activities because nearly everybody uses word processors for specifications. This is done by looking at definitions and then by reference to the widely used NBS/NES (National Building and National Engineering Specifications).

Quite often in many practices the NBS is purchased and used as a basis for the Bill of Quantities or specification clauses on all projects. The close interrelationship between drawings, specification and cost analysis, means that spreadsheets have been found to be of great benefit.

Performance specifications are the only means which a client has to influence or check expenditure and quality standards during progress of the work if he/she has to agree to what is being called for on their behalf. Therefore as many performance requirements as possible should have definitive and measurable criteria, but there is not much point in specifying these unless they are measured and meet a particular standard, e.g. sound insulation of window systems. Furthermore, the writing of specifications needs to be co-ordinated alongside the preparations of drawings — while ideas remain fluid — rather than being left until the drawings are almost complete. Specifications should clearly communicate the project requirements to the contractor and genuinely reflect the design content. As a cautionary note, some word processing packages using standard specification languages can cause and repeat errors unless carefully used.

A performance specification seeks to spell out design intentions, quality and durability expected, levels of thermal performance, sound

insulation and fire protection requirements, provision for piping and conduits,
foreseeable interfacing problems at
package boundaries, the nature and length
and bonding requirements for warranties
or guarantees, etc.

The hierarchy of building performance
specifications which should move from the
general to the particular are: *materials <
components < elements < services <
building*. Also, the required performance
should relate to the expected usage and
should not be pressed beyond the
reasonable accepted standards to meet the
level of financial targets.[4]

Performance specification writing needs
skill and a top-quality technology base.
The US has had a long head start in this
respect. For example, John S. Gero notes
the computer-based specification
production systems which aim to automate
various aspects of specification
production.[5,6] He states

*The specific aims to be achieved in
automating specifications may be
summarized as follows: to produce
better specifications; to produce specifications faster; to produce specifications
cheaper; to release highly skilled professional men from much of the clerical and
mechanical work associated with hand
methods of production . . . SPECS is an
example of a computerized specification
system which attempts to account
directly for the special hierarchical and
relational structure of architectural
specifications. SPECS was developed by
Engineering Computer Internal, Inc.,
Cambridge, Mass., under contract to
Automated Procedures for Consulting*

Fig. 12. Comparison of vector and raster images using analytical computer modelling: (a) series of computer models showing natural light elements of the building that are integrated design components of the physical form; (b) series of computer models and renderings showing natural solid-void features of the building for understanding the massing of the physical forms. The drawings in (a) and (b) show progression from original vector images in a CAD modelling platform to raster images that are saved as TIFF for image manipulation for publishing processes (courtesy of Professor M. Saleh Uddin)

Fig. 12(a)

Wire frame computer model created by Form-Z modelling and Rendering application software. The image is a vectorized model having lines with precise dimensions and directions

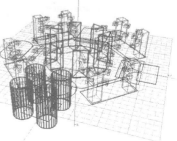

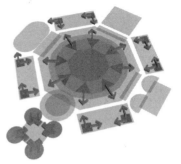

Rendered plan view of the same model

Hidden-line perspective view of the same model

Rendered perspective view from Form-Z brought into Adobe Photoshop software to resize and save as TIFF format for further manipulation. The image became pixelized and changed into raster

Direction of light in a 3D rendered model

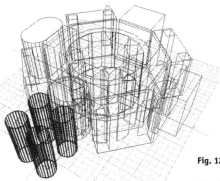

Fig. 12(b)

Wire frame computer model created by Form-Z modelling and Rendering application software. The image is a vectorized model having lines with precise dimensions and directions

The hidden-line vector drawing clarifies the forms in the model

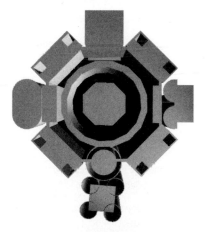

Rendered plan view of the model

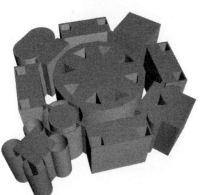

Rendered perspective view of the same model in Form-Z saved as a TIFF file and imported into Adobe Photoshop software to resize for further manipulation. The image became pixelized and changed into raster

Engineers, an engineering computer user's group.[5]

Since the 1950s or earlier, management teams in various projects and organizations have included expert specification writers (sometimes partners or directors) often with their own department. The UK offices have mostly been in a muddle about specifications — sometimes these are left to project leaders, with variable results, or to quantity surveyors who are sometimes given notes and left to do the writing or even just build them into their Bill of Quantities, thereby giving the contractor a lot of leeway on site.

National Building Specification/ National Engineering Specification

The National Building Specification goes a long way towards raising standards of specifications. Because researching, writing and maintaining clear, succinct clauses from scratch for each project is very time consuming, maintained libraries of clauses from the National Building Specification and the National Engineering Specification[7] have gained widespread use for this purpose. The use of computers can help the specification writer save time by assembling packages of text into semi-finished form for final editing and approval. It is usual, however, for specifiers using commercially produced libraries to edit the basic text to

- make it more directly related and relevant to the technical preferences of the practice/office, either by adding special clauses or deleting others

- reduce project specification preparation time by reducing the number of options and alternatives and the amount of information to be inserted

- apply greater control to the technical output of the practice/office, for example by standardizing the choice of proprietary products wherever possible.

These control and refinement aspects of specification constitute elements of consistency. Links with other production information outputs, especially the drawings, are required to ensure logical results and avoid discrepancies and omissions. The software *Specification Manager* ® is an adjunct of the NBS, being a computerized library of NBS specification clauses and guidance notes, plus keyword searching and easy editing. Product information is based on the RIBA Product Selector and interactive accessing of documents such as British Standards from the RIBA Construction Information Service (the CD-ROM of the technical indexes and RIBA Services).

Specification Manager (consisting of a series of computer databases with sophisticated and highly integrated search, management and editing facilities) therefore provides the resources to address these issues and implement procedural models such as CPI (Co-ordinated Project Information). CPI can save time and effort while allowing flexibility to develop creative aspects of design and to achieve consistency and compatibility of output. With a quality management system it provides an audit of project specification and office standardized specification.

Basic components of a specification manager system include

- a drawing

- a specification

- an annotation library for predefined annotations (for example E10 in-situ concrete or E10 ready-mixed concrete or E10 concrete to BS 5328) to be placed on a drawing or to produce reports

- a measurement phraseology library for automated billing.

Specification Manager allows integration with CAD and has now been developed into *Specman*. Specman is one of the first serious AEC IT applications to use the Windows operating system. The question asked is what difference is there between Specman and a good word processor using National Building Specifications? The answer to this question lies in the term 'management'. Specman must cover all the relevant details determined by the specification manager. Where clauses are added or deleted, Specman carries out an audit trail so that the user, time, date and clause are all recorded. As each worksection and clause is being worked on, Specman automatically flags up the relevant guidance, either general to the worksection, specific to the clause, or with reference to documents such as British Standards, and so on.

NEStar and WinSpec are commercial software, available on an annual fee-paying basis and produced under licence from NES (National Engineering Specification) Ltd by Max Fordham and Partners to facilitate the text manipulation involved in preparing a specification. The software allows the user to exploit the essential features of NES, namely

- of enabling practices to keep specifications up-to-date without committing large amounts of manpower to the job

- making the specifier think about every aspect and prompting decisions so there is less likelihood of leaving things out

- ensuring the specifications are short and to the point leaving the contractor with less scope for error, omission or unwanted changes.

References

1. Wyles C. Put Information Technology in its place. *Global Management,* 1992, Management Center Europe.

2. Wilkinson, C. *Information technology in the office*. City and Guilds/Macmillan, London, 1992, 60.

3. Harper D. R. *Building the process and the product*. The Construction Press, London, 1978, 64.

4. *ibid.*

5. Gero J. S. Specifications by computer. *Arch. Sc. Rev.,* **19**, 1, March, 1976, 10–13.

6. Gero J. S. *Computer applications in architecture*. Applied Science Publishers Ltd, London, 1977, 134–172.

7. David J. Words of wisdom: national engineering specification: practice news. *Building Services Journal,* Oct., 1996, 33–34.

Time and project management activities

Introduction

The management and recording of the time spent on projects/jobs by individuals or groups of individuals or organizations are activities which can be monitored for resource planning and management. Project and time management programs allow users to keep track of appointments, the progress of long-term projects, and to indicate the progression of independent tasks. These address the concern of computer systems to streamline workflows and flatten corporate structures, acknowledging that data does not recognize departmental barriers. Alex Reid, RIBA Director-General, points out that business and project management is one of the key areas in which architects can benefit from sound investment in IT, because in such areas computer methods are just so superior to any alternatives.[1]

The use of computers is sweeping through the profession. I am completely convinced that all practices that don't have computers should get them. That doesn't mean they have to have elaborate, expensive systems (they can be quite basic). For large practices working on larger jobs it is no longer possible to avoid investment in IT. For the small practice computers are almost unavoidable because you are so self-reliant. Today you'd find that small businesses of all kinds have personal computers; they just cannot operate without them. Much of the everyday drudgery of running a practice can be removed by using computers, freeing staff for more creative tasks. Three key areas where architects in particular can benefit from sound investment in IT are CAD; information storage and retrieval; and business and project management. In all these areas computer methods are just so superior to any alternatives. Time-saving benefits of CAD are already well charted. Use of information on CD-ROM, on-line databases is another area.

This section discusses management accounting, designed specifically to serve the needs of project managers and to aid decision-making and planning, as distinguished from financial accounting and reporting required by law or government bodies which is covered in the next chapter. At one level, however, examples of management accounting are investment appraisal budgets, profitability analysis and long-range planning, while at the other level examples are time-sheeting, job costing, WIP (work in progress) fee estimating and forecasting, project/resource planning and expense handling, etc.

It is also useful to define and distinguish such terms as DCC — document management, MIS — management information systems (often integrated with financial accounts and needs of balance sheets), and PM — project management (involving project networks, work quantification and method design). In many cases project organization is typified by a continual formation, dissolution and reformation of teams of managers, staff and specialists. It is precisely because clear project co-ordination requires good records and files for future use, and time recording is useful for accounting, that computer systems become relevant.

Appointment schedulers, personal information managers (PIMs)

Computerized appointment schedules closely resemble the paper-based appointment books which are their counterparts. A PIM is useful and powerful because it presents information in a variety of ways: information is recorded and can be viewed as you would in an appointment book, and it can also display the information grouped by person to whom the task descriptions pertain (e.g. Call electrician next Monday about replacing circuit). PTMs may be the best choice if you routinely perform a wide variety of tasks and need to see quick summaries of them, sorted in different ways.

Examples of PIMs are

- Lotus Agenda (Lotus Development Corporation)

- Microsoft Schedule (Microsoft Corporation)

- Instant Recall (Chronologic)

- Sharp Wizard (Sharp Electronics).

Computerized time management involves

- keeping an on-screen diary of appointments, allowing easy editing and creation

- creating and viewing an on-screen to-do list with priorities you assign
- setting an alarm that will warn you of an impending meeting even if you are using another program
- dialling the telephone number of the person's record you are displaying on-screen
- viewing tasks in a variety of ways.

Resource scheduling

The Gantt or bar chart (Fig. 13) is often used as a simple and readily understood means of visual presentation of a construction programme. Bar charts are the simplest form of planning. Quite simply, each activity is represented by a line (bar) whose length is proportional to the time taken to do it. Once all the activities are listed with their respective lines, it is possible to see how long the overall project is going to take or, during the project, monitor progress to see which activities have slipped and, consequently, how long the overall project is going to be delayed. Their limitations are the number of activities which can be plotted physically on a sheet of paper in comprehensible form and the number of interactions, or dependencies, of activities, especially in large projects.

Network analysis methods (CPM — critical path method and PERT — programme evaluation research task or programme evaluation review technique) provide alternative expressions relying upon the basic network plan as a graphical portrayal of the way in which a project should be carried out and the logical sequence of operations.

The many detailed books, as well as British Standards,[2] on PERT and PERT-like systems can be used and are available for consultation. What is important here is the effective use of such methods utilizing computers (to overcome the inflexibility of manual methods) and their sophisticated software in practical applications, thereby making control of projects easier and helping to deal with the sheer volume of data generated. *Project managment with CPM and PERT* is one of the earliest publications to list computer programs and trends of network programs.[3]

CPM was begun in early 1957 by Morgan R. Walker of the Du Pont Engineering Services. He collaborated with James E. Kelley Jr., of Remington Rand to provide a more precise and dynamic model for scheduling purposes. Setting out the graphical network diagram, Walker and Kelley[4] produced a scheduling system which demanded the use of simple, straightforward, arithmetical processes (addition, subtraction) to show that every project could be programmed with at least one sequence of operations, jobs or activities. The result of the calculation is the definition of a critical path through the network diagram, and any of the activities on the critical path which are not completed within their estimated timescale will inevitably cause the overall project to be delayed.

By May 1957 Kelley produced a solution to the question of optimizing the total length of time for a project at minimum cost. By 1958, a plant maintenance shut-down was scheduled by CPM and in March 1959 was effected at Du Pont's works in Louisville, Kentucky. It was claimed that the shut-down period of the chemical plants was reduced from a normal 125 hours to 93 hours using CPM. Fig. 14 shows a CPM example.

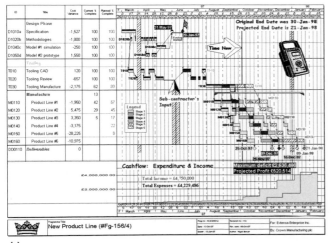

(a)

Fig. 13. Construction project programme planning: (a) Gantt charts; (b) bar charts

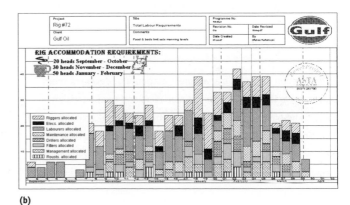

(b)

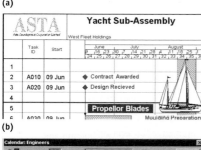

(a)

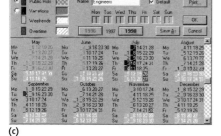

(b)

Fig. 14. (a) Bar charts using PowerProject — activities are drawn directly on screen, using a mouse, in a similar way to drawing plans on paper before performing Critical Path Analysis; (b) charts using PowerProject — information can be conveyed using segmented linked bar chats; (c) calendar by PowerProject — customized and multiple calendars can be created to track the availability of any task or resource; modelling using PowerProject, (d) resource profiles to model the availability and usage of resources, (e) histograms and graphs for individual resources as well as resource groups, (f) displaying of progress as a jagged line to show which tasks are ahead of or behind schedule, (g) multi-level hierarchical project structure for sub-projects and activities, (h) bar charts, text reports, histograms and cash flow graphs can be generated for sub projects, (i) customization of plans

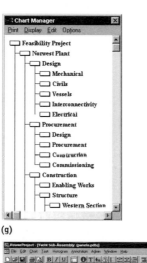

(c)

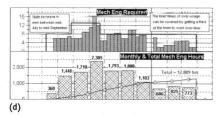

(d)

(e)

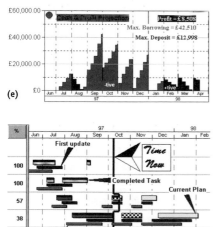

(f)

(g)

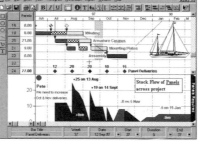

(h)

(h continued)

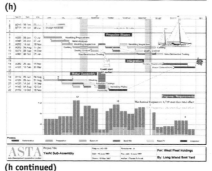

(i)

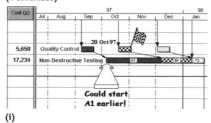

By June 1958, as a result of the collaboration between Waillard Fazar of the Special Projects Office of the US Navy, with D.G. Malcom, J. H. Roseboom and C. E. Clark of Booz Allen and Hamilton of Chicago, Illinois, PERT was developed to produce a method of programme evaluation for the development of fleet ballistic missile weapons. This project involved a very wide range of activities.

PERT was born from a distillation of bar charts, milestone reporting systems and line-of-balance management techniques. It pays particular attention to controlling the time element and employs statistical analysis to evaluate the probability of meeting target dates throughout the duration of the project. The production of progress reports is another major aspect of PERT systems. The PERT method of scheduling was effective and it is now credited with making a considerable contribution to the completion of the Polaris missile programme ahead of time, involving some 3000 contractors and agencies.

The RIBA handbook (1965) sums up all such methods developed on the basis of wartime experience after 1939, when it became abundantly clear that production could be seriously delayed when certain parts were in short supply and subject to long delivery periods

PERT: developed for the US Navy 1957/58 originally concerned more with 'events' than activities also uses three time estimates for each activity — optimistic, likely, and pessimistic. CPM: similar but more concerned with activities and uses one time estimate for each.[5]

Time, project and construction management

With all these methods (CPM, PERT, etc.) the network is used to portray in graphical form the logic of the construction plan. It is really a project graph — a means of representing a plan so that it displays clearly the series of operations which must follow in order to complete the project. It also shows clearly their interrelationship and the interdependence of one with another. One major advantage of the method is that the planning phase of the programming operation may be divorced entirely from the scheduling or the estimation of time durations (for example hourly, daily, weekly, monthly, etc.) for the project.

Therefore the network defines its programmed activities by grouping tasks under headings which are based on design elements (i.e. ground floor slab, columns, first floor slab, rainwater pipes, etc.), whereas the processes of construction make for a better matrix, and these are measured in terms of activities. Construction activity is shown to have three factors

(a)   content — the individual tasks

(b)   structure — one activity is inevitably linked with others, dependence and interdependence can involve both sequence and concurrence

(c)   tuning motion when and at what speed, which may depend upon energy, or on following on required preparatory work.

Three basic methods of preparing a diagram are

(a)   activity-on-the-arrow system

(b)   activity-on-the-node system

(c)   event or milestone system (e.g. PERT).

Computers can deal with larger and larger networks in more detail, allowing many variations and they have the ability to address diverse requirements. Examples include whether or not to split an activity and work within resource restraints to produce aggregation summaries within trades, or produce total aggregation figures. These networks are effective for simple repetitive work. D. R. Harper shows a CPM program by BRE (Building Research Establishment) and Northamptonshire County Council for Daventry County Primary School in the 1960s.[6] He points out that problems of CPM time control are

- requirements of extensive site experience. This includes familiarity with the use of networks and resource scheduling principles

- building operations are only very roughly and arbitrarily consecutive, such that many can proceed concurrently if distributed over the site of the project

- updates of a network require as much skill as that required to create the original programme

- the CPM programme is very difficult to digest for anyone who has not had a hand in preparing it

- if not site based or located, programmes are nearly always grossly over-optimistic

- an ideal programme could show a direct relationship to costs at a date line reached if there were not two very difficult wayward factors: (a) a programme can hardly show the necessary time which may regularly be required to achieve quality/standard, and (b) it cannot anticipate the success or failure of forward delivery times so that arrivals are precisely timed. Neither of these factors are very amenable to forward planning.[7]

Typical project management software, including scheduling programs covering all these methods, is listed below

- MacProject II® (Claris Corporation)

- Microsoft Project (Microsoft Corporation)

- Harvard Project Manager (Software Publishing Corporation)

- Quickdex (Casady and Greene)

- SuperProject® (Computer Associates)

- ARTEMIS (Lucas Management Systems)

- CRESTA (K & H Project Systems Ltd)

- OPEN (Welcom Software Technology International).

MacProject II can be used to produce a master critical path analysis programme showing the overall project phases. Such project management programs go beyond appointment schedules or PIMs in that they include tools for viewing a project's schedule (in the form of an on-screen Gantt chart or time line) and for tracking project expenses. Another package used in critical path analysis is PowerProject, running on Windows.

PowerProject was developed and marketed by Asta Development (based in Thame, Oxfordshire). It was reported in an independent survey by David Bordolli that PowerProject is the most widely used project management software in the construction industry. There are regular updates included in the annual support contract cost of £160 per annum. The software is said to have a very direct approach which by-passes all the theory and deals simply with the bars of the bar chart themselves. Once arrows are drawn with the mouse to connect the bars, PowerProject does all the hard work to identify the critical path. It can easily provide month-by-month cash flow breakdown for the whole job. Histograms can be rapidly set up with suitable axes, allowing graphical presentation and information to be displayed, that is income and expenditure — including overspend and underspend. This allows rapid production of a graph based on the bar chart timing and cost allocation.

PowerProject easily allows fixed costs to be allocated to individual bars, as annotation facilities within PowerProject allow the printing of cost on the bar chart in the font and size required without too much difficulty, and also allows the same annotation to be automatically placed on each of the other bars.

PowerProject Professional benefits from the fact that critical path analysis was one of the earliest and most powerful applications available on microcomputer to design professionals. This was because the theoretical basis was already well established by 1980. However, the reiterative nature of calculations made them very inflexible to do

long-hand, complicated by the fact that each time minor adjustments were made the whole process had to start from scratch.

In general, with computer charting methods there is flexibility and a reduced risk of logical and other errors being introduced, especially with most real projects having over 100 activities. The programs offer opportunities beyond running time analysis and printing out the results (i.e. running projects using the earliest possible dates with little regard for the consequences of resource constraints).

The arrival of the microcomputer was just what the academic specialists had been waiting for and soon a number of increasingly sophisticated products were developed — many of which still survive, including Hornet from Claremont Controls (originally developed by architects), MicroPlanner (an early Apple Mac product), Artemis, SuperProject and PMW. As with all applications, established software tends to drift up-market as it becomes more sophisticated, leaving behind the lower end to new companies and new products which also take advantage of changes in technology, such as Primavea which uses HTML-based web reporting.

Computer-aided facilities management (CAFM)

Facilities management is nothing new; many organizations have long contracted-out their catering, publicity or security services as the most cost-effective way of managing these essential aspects. Many systems for control or management of such divergent spheres as assets and cabling, or operation and maintenance of facilities, are based on such principles, with critical functions of management control, collection and storage, analysis and prediction.

Figure 15 shows a useful classification according to activity types which allows a breakdown into relevant constituent elements. The relevant primary data groupings are: property asset particulars; costs and income; conditions and suitability; premises strategy and standards; and work procurement and administration. These contain a range of detail which is dependent on whether the interest is from a strategic or a day-to-day management perspective.

The main idea behind computer-aided facilities management (CAFM) systems is that they are used not only to automate FM tasks but to integrate different aspects of facilities management, thereby providing a management information system. Typical software packages consist

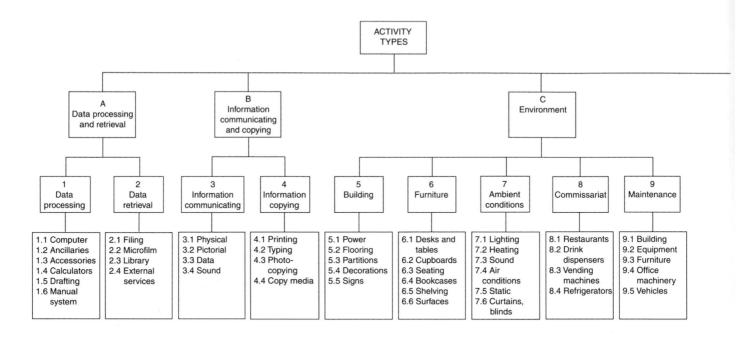

Fig. 15. Activity-types organizational index

of a CAD package linked to a database, with the advantage that drawing and database information are always synchronized so that changes to the drawing are reflected in the database and vice versa. This is more useful for applications where drawings are essential to the operation, such as space planning. Two main problems concern the level of integration between CAD and database and the complexity of the CAD system in general, which conflict with the need to develop such features as expandability, openness, and system architecture to allow data to be moved around the organization and from one system to another. Development of systems where drawing objects and database records are stored as one element would be of immense help in overcoming these deficiencies.

Typical CAFM systems include

- Archibus/FM supplied by Archibus, runs under Windows and combines a relational database and AutoCAD but can be used purely as a database system. The database can run with either SQL or Oracle. Applications modules provided are space management, asset management, planned maintenance,

property and lease management, telecoms.

- AutoFM Desktop supplied by Decision Graphics consists of self-contained modules, e.g. space planning, room booking, etc.

- IFM consists of space, asset, lease and maintenance management systems by Intergraph.

- Drawbase supplied by Ground Modelling Systems incorporates a spreadsheet database and enables the production of information for an accommodation strategy, planned maintenance and space utilization. In contrast to other PC-based CAFM products, its flat database is integral to the system and it does not use AutoCAD but it can link to external relational databases, and also import and export AutoCAD DWG files.

- Mountain Top supplied by Acunet is a CAD-based system for UNIX workstations, combining drawings and databases such as Oracle, Ingres and Sybase. Typically marketed as a network management system, it also provides functions to manage space planning and space charging. Its greatest strength is the

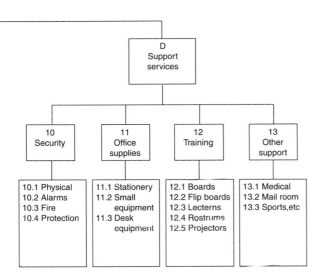

scenario modelling, as well as assets to be physically tracked, accounted for and planned in a flexible manner; critical date tracking (AP/AR); landlord versus tenant services and financial support information, and full range tracking and planning for PPM, materials handling and failure analysis. The new BIFM (British Institute of Facilities Management) states the following about the role of computers:

Computer aided design systems (CAD), though not essential to good space planning, can be a valuable tool. They may be used simply for drawing layouts, or for experimenting with 'what if?' situations and recording real change. Computer aided facilities management systems (CAFM), as the name implies, are more complex tools intended to assist space planning, as well as with the broader management process. In both cases, the systems are only as good as the resources available to keep them in use and up-to-date.[8]

References

1. *Building*, 9 June, 1995, 53.

2. British Standards Institute.
    BS 6046: Part 1: 1984. *The use of network techniques in project management: guide to the use of management, planning, review and reporting procedures.*
    BS 6046: Part 2: 1992. *The use of network techniques in project management: guide to the use of graphical and estimating techniques.*
    BS 6046: Part 3: 1992. *The use of network techniques in project management: guide to the use of computers.*
    BS 6046: Part 4: 1992. *The use of network techniques in project management: guide to resource analysis and cost control.*

3. Modier J. J. and Phillips C. R. *Project management with CPM and PERT.* Reinhold, New York, 1970, 109.

4. Kelley E. J. Jr and Walker R. M. Critical-path planning and scheduling. *Proc. Eastern Joint Computer Conf.*, October 1959, in Davis E. W. (ed.) *Project management: techniques, applications and management issues.* Industrial Engineering and Management Press, Institute of Industrial Engineers, 1983, 113–118.

5. RIBA Handbook. *Architectural practice and management.* London, 1991.

6. Harper D. R. *Building, the process and the product.* The Construction Press, London, 1978, 97.

7. *ibid.*, 356–58.

8. Cochrane McGregor and Associates. *Best practice guide: space planning.* For the British Institute of Facilities Management, London, 1996.

ability to link running database and graphics seamlessly. Also, it works with a standard programming language allowing complete applications to be built on top of the system.

- QFM supplied by Service Works is a database system for PCs and client/servers with applications for planned maintenance, helpdesk, lease management.

- Planet G5 supplied by FDM has applications for planned maintenance, helpdesk, purchasing and stock control. Works with a Paradox database on standalone PCs or small PC networks, or on a Sybase database.

- Concept 500 supplied by FSI can be used with databases such as Oracle, SQL and Sybase, which enable it to function as a multi-site system on LANs and WANs. It can link to CAD and graphics but does not provide full integration. Applications include planned maintenance, helpdesk, stock control and purchase ordering.

Such programs allow: comprehensive space planning, including personnel and space tracking; charge-back by floor area, and 'what if'

Financial accounting and modelling

Introduction

This chapter is about office automation in terms of accounting/book-keeping, financial modelling and activities involving numerical analysis, including structural analysis and spreadsheets. Spreadsheets are the second most common application of PCs after word processing. Also spreadsheets are effectively the computer software variant of the accountants' old paper ledger, and, because financial accounting is required by law, they have to obey many rules and strive for objectivity.

Spreadsheets

Spreadsheets are based on the simple concept of a sheet of paper which is spread out and then filled with a table or matrix of rows (referred to by numbers) and columns (referred to by letters). Figure 16 is a graphical representation of a typical spreadsheet. A spreadsheet allows text, numbers and formulae to be entered in a matrix of 'cells' (the node or intersection of rows and columns) and then manipulated for analysis of performance or 'what if' scenarios. They are quite useful despite their simplicity, and can be applied to most tasks requiring the presentation and manipulation of numerical data.

Advantages of using spreadsheets include

*   entering and editing of data and labels, thereby gathering all information in one matrix and offering minimal re-keying, avoiding the need for proofreading and the consequences of making errors

*   entering formulae such that, once put in and tested, they need not be altered. Existing formulae can be copied to other cells and printed out

*   performing 'what-if' analysis, allowing projections to be carried out easily because recalculation and choosing formats is easy, as cells can be set to the size required (e.g. A4 size page). They can be viewed on screen and printed out as defined sections, and by condensing the printing more of the data can be shown on a page

*   graphing, so that data can easily be translated into graphs or charts and included in a word processing report. Information can be exported/imported to/from other packages and combined in a report

*   the ability to produce a generic or standard outline worksheet which can act as a template for a variety of tasks.

Spreadsheets generated by computer are referred to as 'electronic spreadsheets' and these offer certain advantages, e.g. they avoid the lengthy calculations by hand which are notoriously prone to calculation errors, especially if the process has to be repeated. Cells can either contain data, a label, or a mathematical or logical formula based upon other cell references. By putting a small step in a calculation in successive cells some useful models can be built up. For example, a spreadsheet can be used in marketing for pricing, marginal profitability, sales productivity, investment analysis and distribution.

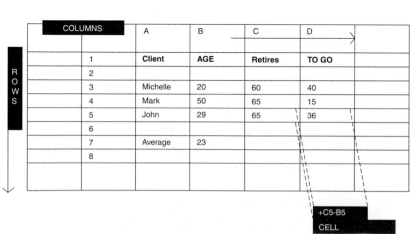

COLUMNS		A	B	C	D	
	1	Client	AGE	Retires	TO GO	
	2					
	3	Michelle	20	60	40	
	4	Mark	50	65	15	
	5	John	29	65	36	
	6					
	7	Average	23			
	8					

+C5-B5

CELL

Fig. 16. Typical spreadsheet

As simple devices, spreadsheets allow financial and numerical presentation applications, and 'what-if' revisions of the table. For example, in sales forecasts for strategic planning and market research, variables such as raw material costs and interest rate values could be assessed by asking 'what if raw materials costs doubled, what would be the result?' In other words, the ease of answering such questions is the basic function of spreadsheets.

The irritating part about most 'what if' analyses as applied to forecasting is that they do not really answer the question you are asking — or trying to ask. The situation is often that you know the likely range of values for the input cells and want to know the corresponding likely range of values for the output cells. However, this can also be a basic drawback, since over-optimism might well lead to manipulation of the data to obtain a hoped-for rather than a genuine iteration. The other problem that can arise is that entering an incorrect formula in a cell can have disastrous consequences. This may not always be obvious with initial iterations.

The original spreadsheet, VisiCalc by Dan Bricklin (now president of Software Garden) of Arts Software, was launched in 1978 and remains the basic model for all spreadsheets. It created an automated way of working with spreadsheets, opening the door to new ways of thinking about solving problems with technology. However, typical spreadsheet software packages which have since been developed include the following, some of which also offer a range of basic statistical functions

- Lotus 1-2-3 (Lotus Development Corporation)
- Quattro Pro (Borland International)
- Excel (Microsoft Corporation)
- SuperCalc (Computer Associates).

Most spreadsheets accomplish the tasks of providing totals, averages, data deviations and net present values via '@ functions' — e.g. '@Sum'; '@AVG', etc. — @Sum (B3, B4 ...B5) being the formula for totalling entries in cells B3, B4 and B5. If one numerical entry is altered by the user, all formulae dependent on its value will also change. This aspect, together with the extremely easy manipulation, is what makes spreadsheets particularly powerful.

In his book, S. R. Davis shows various examples of the applications of spreadsheets.[1] Almost all practices use spreadsheets

anyway, especially when talking to clients about cost programming for their proposed development and developing a model to demonstrate the impact of design options and opportunities in their business. Increasingly, spreadsheets are being used not only for financial analysis but also for scheduling, database management and graphics presentations.

Financial accounting

While spreadsheets or databases are often used to operate account systems, dedicated accounts software is also available. Business accounts are essential both to comply with the law and to run the business, not least because they record transactions with debtors and creditors. The use of computers makes possible effective control of funds coming in and going out of an organization. This is particularly true if the vast majority of accounting data is always processed using the same method, i.e. to show control of income and expenditure within the business and to establish and monitor its financial targets.

The advantages of having a computerized accounting function have been well documented and these include the following.

- Addition and entries should be more accurate (because automatic recalculation leads to error prevention), in particular where there are very large quantities of data to be entered in the accounts.

- Reports can be produced more quickly and as frequently as necessary; work which would normally be laborious and take weeks to do if done manually and would therefore probably be impractical because of the time and cost involved. These include the Trial Balance (a list of debit and credit balances categorized by account from the nominal ledger); the Transaction Report (a full list of transactions which may be used to check for errors or as an audit trail by an external auditor), the Profit and Loss Account (a statement of trading performance of the business over a given period) and the Balance Sheet (a statement of the business's assets and liabilities at a particular point in time).

The main issue for managers of organizations is that although they know technology can help make them become more competitive, they are often constrained by outdated *legacy systems*. Increasing globalization requires that companies standardize on a common software platform. Internally, firms need to implement new systems to enable cost reduction and/or improve client service by establishing more efficient processes. There is also the drive to provide managers with better information to enable them to run the business more effectively.

Computer programs for financial accounting invariably follow the system of dividing different ledgers into nominal sales and purchases and ledgers. This is for historical reasons — the first companies to use computers were large organizations whose manual records were already subdivided into the different accounting ledgers in which transactions are recorded. Simon Edwards, director of UK operations at System Union, explains that

When the majority of computerized accounting systems were introduced, they were structured into three separate ledgers covering receivable, accounts payable and general ledgers. Often companies bought different ledgers from different software suppliers. Integration was not even considered a requirement.[2]

There is a wide diversity of commercial systems, reflecting the wide variety of users, but the main areas which are commonly computerized are as follows.

- Purchasing Ledger (also known as Creditors or Buyers, or in the USA as Accounts Payable) — which deals with the purchasing of stock, i.e. management of goods (in particular, trading stock) by a business. It gives information on suppliers, e.g. how much each is owed.

- Nominal and General Ledgers — which bring together the various totals from the other ledgers under classified headings for financial analysis up to, and sometimes including, final accounts (e.g. Profit and Loss Account Statement and Balance Sheet) so that the current financial position of the business can be assessed.

- Payroll — which facilitates the calculation of wages so that payslips, cheques or credit transfers used for payment, coin analyses and other related information can be produced for employees.

- Sales Ledger (known as Accounts Receivable in the USA) and associated Sales Ordering, Invoicing, and Sales Analysis — all of which deal with the sale of goods (either manufactured or from stock, or from services) by a business. The Sales Ledger gives information on customers, e.g. how much each owes.

- Stock Control (or Inventory) — which deal with additions to, removals from, and the storing of (and therefore management of) trading stock, e.g. stock items such as raw materials and finished goods held by a business.

- Production Control and Job Costing — which deal with the production processes and their costs and these range from large one-off items (such as ships, roads and buildings), batch items (such as nuts and bolts, books) to bulk or flow-processing items (such as ice-cream, milk, petrol, etc.).[3]

Most commercial accounting software is now available as integrated packages, which means that although it is possible to buy a package covering just one area (e.g. sales, nominal or purchases), when all the packages are brought together the entries in one ledger will automatically update accounts in another ledger package. Figure 17 shows the main interfaces between the major commercial systems.[4]

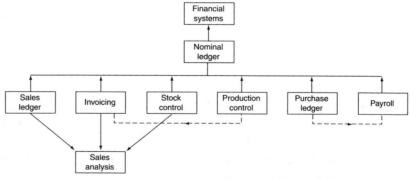

Fig. 17. The main interfaces between the major commercial systems when integrated

This is convenient and benefits most companies except for those who may only have a handful of suppliers (e.g. timber merchants), or pay for all purchases in cash (e.g. scrap metal dealers, antique dealers), or those companies who do not purchase raw materials (e.g. computer bureau). In these circumstances a purchase ledger package may be a waste of time, but a sales ledger package is a 'must'. Other companies (including a lot of very large businesses) have no credit sales (e.g. supermarkets), or sell to a limited number of organizations (e.g. shipbuilders or airplane manufacturers).

For these companies, conversely, a sales ledger package may be a waste of time but a purchase ledger package is essential, given that they buy very large quantities of raw materials over a wide range of stock. Some organizations have very few purchases or sales on credit, and therefore do not need sales or purchases ledger (e.g. small shops, football clubs, theatres, etc.), and for these firms, a nominal ledger package may often suffice. Examples of accounting programs include

- Sage (Sage Software). Suite of programs comprising 12 separate business accounting modules plus the system manager and EIS (Executive Information System)

- Quicken (Intuit)

- MacInTax (Softview)

- Peachtree Complete III (Peachtree)

- Mind Your Own Business (Teleware)

- WinAccs (by Pinstripe Software — the first Windows accounting software).

When selecting packaged business applications, it is important to consider both the product and the vendor. Buying from larger vendors with a track record of delivery carries low risk, and has the added advantage that these have a large installed user-base which attracts third-party alliance. Large vendors and their products include the following.

- SAP with R/3 product (probably the most functionally rich integrated Enterprise Resource Planning — ERP); the clear leader in the large corporate market and dominant in some industries like manufacturing.

- Oracle — a major software player, with the leading client/server database. It is also a leading player in its own right in the business applications market.

- PeopleSoft — its human resources product is the market leader and its financial product is functionally rich and easy to use.

- JD Edwards — a large ERP vendor that is financially strong, and has become established in the AS/400 market. Another of its products is OneWorld which allows it to enter into the client/server market.

- Baan is used particularly in the manufacturing sector. Generally the system is very flexible and its integrated tools cut implementation times.

- Lawson has a strong reputation in the AS/400 mid-range market with some excellent features such as the Activity-Based Costing module and Drill-Around reporting.

- SSA's BPCS product is an integrated manufacturing, supply chain and financial product.

- Dodge has been very successful in selling and delivering its Open Series GL Data Warehouse product to the investment banking sector.

- Hyperion is the market leader in the financial consolidation market.

These accounting programs allow management of invoicing and other activities.

New types of integrated package, such as Symphony and Framework, generally include a word processor, spreadsheet, database and a graphics package so that data from one element (e.g. a database) can be integrated with data from another element (e.g. a word processor). The following combinations are frequently offered: word processing with database, spreadsheet and communications; spreadsheet with database, graphics and time planner; and word processing with diary, calculator, clock and graphics. Such packages offer certain advantages

(a) only one program need be purchased to fulfil the firm's requirements, thus saving costs. Such programs are often provided free with a computer system and are often suitable for use by a novice, as they are normally user-friendly and provide sufficient facilities for a beginner and/or an organization who do not need a sophisticated system

(*b*)   once one section is mastered by a user, it is easy to master the others, giving further time and cost savings.

They are integrated but should not be confused with integrated accounts packages which will invariably contain more than just sales, purchase and nominal ledgers, as previously described, but may also include other business packages such as

- sales order processing — receives customers' orders and initiates the process of order fulfilment

- stock control — records stock movements and controls stock levels

- invoicing — the production of invoices requesting payment from customers for goods or services supplied.

Billing for time and expenses

The largest single item of expenditure in an architectural/ engineering practice is staff salaries and these are directly related to staff time. Thus the control of staff time is of considerable importance for the measurement of office productivity. In addition, programming of individual projects allows the control of time and, through time, costs. Furthermore, as more and more architects and engineers charge by the hour, increasingly they face the problem of how to keep track of the time employees expend on each account, and how to bill clients correctly.

Special-purpose programs are useful in this context, such as Timeslips III (Timeslips), a market-leader, or MedPac (Syscon Computers) designed for medical practices, or TimePiece Legal (Impact Software Productions) for legal practices. Such programs combine database management with event timing, billing, accounts receivable tracking, and other features. They offer two great benefits

(*a*)   the time saved in billing clients

(*b*)   the increased access to fundamental information about the practice's performance.

A time-billing program is not an accounting package in that it can perform the following tasks.

- Record each billable event as it occurs or later, showing who performed what for whom, and for how long, thus allowing billing by person, by client or by the activity. It can record not only expenses (such as mileage, FAX charges, copying, etc.), but also timed activities, so that you can set rates, charge sales or VAT if applicable but also incorporate descriptive notes that identify and explain the activity, add new clients, employees and activities on the fly. It can also sort, organize and search records about specific time-related events (indicating who, what, for whom, and when).

- Time activities as they occur by using an optional terminate-and-stay-resident (TSR) mode, especially when billing for telephone advice when simply pressing the key that displays the time-billing program will automatically time the call accurately.

- Bill at a fixed rate (for example £125 per specific task) or at a timed rate (for example £50 per hour), and then compile all records at the end of each billing cycle and print itemized bills. With mailing facilities to clients, billing goes out quickly and by maintaining a client's account, payments and other transactions can be posted on it.

- Generate reports (data prints) that show which clients have not paid their bills on time and thereby allow overdue accounts to be tracked.

Examples of such programs include Timesheet Expert for Windows (with excerpts shown in Fig. 18), Corporate Expense Management System (CEMS) and BookIt.

A typical time-tracking program example, Timesheet Expert for Windows (by Atlantic EC Ltd, Bradford), allows hours to be logged against projects or jobs and activities, calculating the appropriate cost and charge information, and tracking expenses and bonus information. Its week-at-a glance electronic timesheet offers a comprehensive record of all current tasks, organized by project and client for each employee — with up to nine levels of detail. Hours worked, expenses and notes are easily input, either by individual members of the workgroup via their own PCs or batch-entered by a single user.

ITIM Systems has a Corporate Expense Management System (CEMS) designed to help companies and individuals account for, manage and reimburse travel and entertainment expenditure. The system is ideal for those people who hate doing expenses and leave

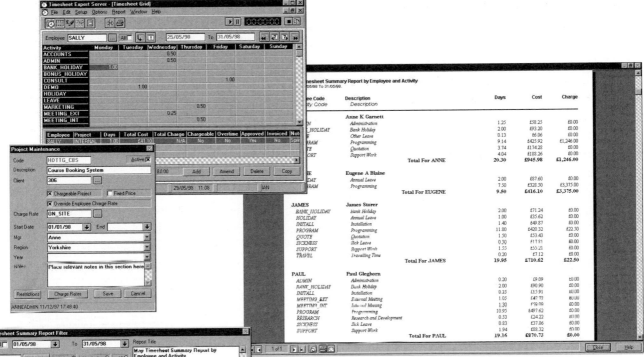

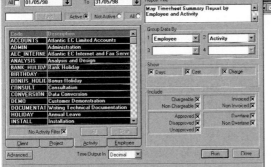

Fig. 18. Excerpts from Timesheet Expert for Windows (courtesy of Atlantic EC Ltd)

it to the last minute. It provides a page on the PC with all the headings laid out, so all that has to be done is to fill it in. The system then handles tax and VAT details.

Another software program, BookIt, supplied by Clandestine Software, seeks to add value to the bottom line by streamlining the whole process of business travel and expenses, making it more efficient and better managed. It takes care of every stage of business travel, from concept to completion. Instead of waiting for the traveller to return with a pocket full of receipts, the software is designed to enable companies to tackle the issue proactively. BookIt will create the travel itinerary at the outset, taking both personal preference and company policy into account when planning the trip.

All expenditure is itemized and classified, including personal expense claims and expenses which need to be charged on to a client or another company department.

References

1. Davis S. R. *Spreadsheets in structural design*. Longman, 1995.

2. Newing R. Accounting software — new ways to make it all add up. *Financial Director,* June, 1995, 45.

3. Dodd J. F. *Practical computerized accounting systems*. NCC Blackwell, Oxford, 1992, 23.

4. *ibid.* 28–9.

Database activities

Introduction

This chapter is about databases and their management. Databases are generally used where complex data storage or manipulation is required, and database PC software is the third most common application after word processing and spreadsheets. Databases are useful to improve the organization and procedures for both architects and engineers, with applications varying from drawing retrieval to client information and space inventories.

Databases

As an elaborate filing system, a database can be compared to a set of index cards in a box that holds information on names (defined by company/organization names, etc.) and addresses (defined by street, city, postcode, etc.) of clients. This card index analogy is illustrated by Fig. 19. Information is then translated into a database table held in a computer database management system. The data held originally on one index card is known as a row or record. The different items (e.g. company name) originally on the index card are known as the columns or fields. Every field must have a field name (e.g. city). Each time the database is used, the computer has to be told which field names to access.

Adding, deleting, changing or sorting this information is 'managing' the database. A database management system (DBMS) is one that enables these tasks to be carried out, and then prints the results in the desired format. Databases can

- store information in an organized manner and then allow access to the data in a variety of ways by enabling a search to be made through the data according to various criteria (cross-referencing or by offering choices)

- offer the facility to sort data alphabetically and numerically, providing lists (reports) on the screen or printed out either completely or partially; they can be menu-driven or programmed so that several functions can be carried out automatically, or calculations performed.

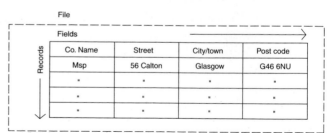

A database table uses three basic elements (files contain records which in turn contain fields)

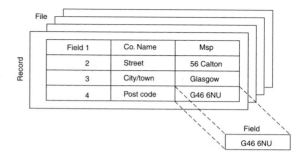

The entire database file corresponds to the set of index cards in a filing cabinet

Fig. 19. The database 'card index' analogy

Database applications can cover student, client or customer records, research surveys, library, property details for an estate agent (e.g. to record for each property the price, number and size of rooms, type of accommodation, area of garden or land) and so on. The categories of databases are

(a) flat-file (the computer equivalent of a card index, but can only handle one file at a time — ideal for storing names and addresses because it is simple, cheap and easy to use)

(b) relational (stores information in such a way that linkages — relationships — can be made between different collections of data, and all of them can be searched together to assemble a new file).

A true relational database stores only raw data, and unlike other types, the links do not have to be set up in advance of input. A query language, which is very easy to understand, is then used for this purpose. Operations (using a range of relational operators based on relational algebra) on relations constitute 'cutting and pasting' exercises to produce new required relations. SQL (Structured Query Language), developed by IBM and supported by ANSI (American National Standards Institute), provides a 'command language' in software packages available for such operations.

It is a fourth generation language (compare procedural language) and to use it there is no need to describe the database, to know how to find the values in SELECT clause, or how to join tables — the DBMS software should do all this for you, including optimizing the procedure used to ensure the best possible access. It is a multi-purpose database language that can be used to define as well as to manipulate and retrieve data. The most frequently available operators include PROJECT, JOIN and SELECT. SQL includes DDL (Data Description Language) designed to define the database, DML (Data Manipulation Language) designed to change or retrieve values in a database, and Query language designed to allow 'on-demand' queries by users.

Database fields may be of two types

(a) numeric (fields used to hold numbers)

(b) alphanumeric or text-fields (fields used to hold text entries). Alphanumeric field length is limited to a maximum of 255 characters, while 20 to 30 is common.

Compared to card-based filing systems, computer databases can easily be regularly 'backed-up' (i.e. copied to other computers or safe data stores). In general in considering data storage there are two options.

1. Centralized (which offers the best solution when you must ensure that all users see exactly the same data).

2. Distributed DBMS (which offers the best solution when most updates are local, when you need only infrequent access to data at other sites, and when you have few updates that span several multiple sites. Even with the advent of parallel database technology, smaller databases are still more responsive and easier to administer).

A WAN connection characterizing distributed DBMS must typically trade off speed and availability for cost. If all queries or transactions must go through the same database, the users are vulnerable to network problems, network delays and system failures. Partitioning the data and using distributed database technology to query it may be the solution to overcome problems of centralized data and distributed DBMS.

Important references must be made to two concepts — data replication and data warehousing. Data replication encompasses a spectrum of technology and techniques for creating and maintaining copies of data at multiple sites while data warehousing is the application of replication and other technologies to bring data from a variety of sources into one or more collections that are designed to improve information access. Issues to be dealt with include proper database design and semantic heterogeneity. Data warehousing is the next pursuer for the SQL server group — in particular the set-up and maintenance of metadata, data warehouse population and administration.

Apart from considering the likely uses of DBMS and the avoidance of input errors, other important decisions to be made when setting up a DBMS include that of the number of characters (bytes) that are to be allowed for each column or field. The length should be kept constant so that the same amount of memory (e.g. 35 characters) is taken up. DBMSs offer major advantages in that they can hold/handle an immense amount of information (e.g. two billion records or index cards for a small town). They enable the database to be searched according to various criteria and will find selected fields or columns very quickly, automatically and accurately. Table 5 gives an example.

Table 5. Searching the database

Symbol	Definition	Example
=	equal to	'= Michael' would find all records naming Michael in that specified field
<	less than	'< 20' all records with a value of under 20
>	greater than	'> 20' all records with a value greater than 20

The computer can simply be instructed to print out letters addressed to those names in the field, e.g. 'owing > limit', which has obvious advantages for credit control. They can be used to handle anything which can be expressed in their basic format — inventories, personnel records and invoicing, stock control, customer records, supplier information and accounts ledgers. Mainframe DBMS programs have included

- Adabas D®
- Datacom®
- IDMS®
- MS®
- DB2®

and for small PCs, programs have included

- Ashton Tate's dBASE® (good report generation)
- R:base® by Microrim, Inc. (good for user-defined rules for data entry)
- Dataease®
- Paradox® (good all round DBMS)
- Foxbase +® (fast)
- Oracle
- Informix.

Oracle, Sybase and Informix are now the three leading relational database companies. There has also been the acquisition of Fox Pro by Microsoft and the development of Access as the office database, with a natural upsizing progression to the SQL Server RDBMS. Another important development has been Borland's takeover of Ashton-Tate.

Document management systems (drawing office management databases)

Before computers became commonplace, architectural and engineering drawings were prepared manually on drafting boards and stored in files and vault rooms. People in practices generally knew who worked on what, where information was located and when projects were completed. So keeping track of design information was fairly straightforward.

All this has changed with the increasing use of CAD. Data is stored on disk drives scattered throughout an organization, under all sorts of different file types and naming conventions. Retrieving information often requires expert familiarity with system commands using arcane syntax. Difficulties are compounded by the growing complexity of CAD models, which may include 3D geometry and associated 2D drawings as well as assembly information, analysis results, numerical control (NC) instructions and metadata such as material type and surface finish. For a practice with 50–60 live projects at any one time, organizing and keeping track of drawings and paperwork is essential. Cue databases or document management systems (Fig. 20).

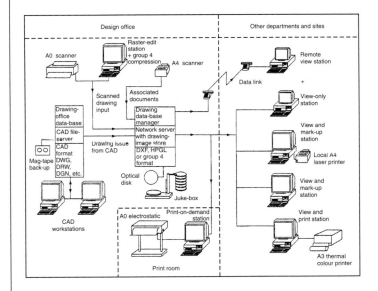

Fig. 20. Typical electronic system for storage and retrieval of drawings and documents (courtesy of Westward Ltd)

Document management systems have grown out of the financial and utilities sectors — banks and building societies, water companies, NHS Trusts — organizations with hundreds of staff, thousands of customers and millions of transactions. They have been using electronic systems to keep track of paperwork and computer-generated files for several years now. In the past, document management users have written their own programs. Increasingly, however, software companies are designing programs that can be bought off-the-shelf for a cost ranging from £200 to £10 000. Common features of these programs are

- checks who is using which drawing or written file on the computer system
- knows which is the most up-to-date version of all documents on the system
- records all documents sent out
- records all documents received
- allows you to send documents internally around the office.

Document management systems can therefore offer two basic functions

(*a*)   check documents in and out

(*b*)   organize workflow.

This means that a central database 'issues' all drawings and paperwork like a librarian would. It knows at all times who has which document and what version of the document is being worked on. The second feature, workflow, keeps a record of when all documents are sent both within and outside the office. It also enables documents to be sent automatically with instructions within the office.

Typically, in a practice every job is given an individual electronic register. The job architects/engineers consult the register via networked desktop personal computers. The register charts drawing schedules, document history and logs the documents that are sent out (both internal and external). The system can also create instructions and can log defects and delays. Everyone working on the project has access to this information and is responsible for updating it, with the feature that it cannot be manipulated at a later date.

Databases for drawing retrieval show more variation and customization than any other application. A large number of packages are drawing office management databases with numbering, approvals and time-logging features. To control large projects, a drawing-tracking database is necessary with issue-details, return-times and alert features.

In both cases change control and dependency between documents in a set are useful features. For process-control, links

to the plant-management and maintenance databases would be more appropriate, and in facilities management, space-calculation, cable records and asset-tracking are required. However, the basic requirement of a database is the ability to call the viewer program with the path-name of the image-file. Closer linkage can be added to produce either simpler entry of known drawing numbers or to produce a more seamless user interface. Document management systems have included

- *Auto-Base* from Cyco
- *Active Data Manager* from ADMIT Systems
- *ReCAD* from Excel
- *Enhancement, Workcenter* from Aprotec
- *CADman* from SMC
- *Cad-Capture* from Cad-Capture
- *EDMS*
- *Document Tracking* for Windows
- *Timesheet Expert Server* from Atlantic EC
- *Document Tracking* for Windows from Micro Planning
- *Keyfile* from Mass Systems
- *Team Mate '96* from Bentley Systems.

In general, an architect database containing design information of a particular project allows any changes to the dimensional information to be reflected in views/images produced by computer modelling. A major advantage of an electronic retrieval system managed by a DBMS over paper is a central store of drawings and documents with on-line network for many users. The location of the drawing (i.e. its file path-name) is stored in the database along with other information on the drawing's creation status, related parts and so on. The drawing image can be stored on any network file-server.

Generally, document management is an electronic system of filing both drawings and paperwork, automatically controlling who has access to these and keeping abreast of revisions. The area of document management can be separated into distinct areas of

- creation concerning PC and networks, mainframe computer output and desktop publishing/graphics studio

- document reproduction covering convenience copiers and fax, a centralized reproduction department, a data centre, printing facilities and offset litho reproduction
- document distribution concerning intra-company mail (i.e. internal distribution) and production mail (i.e. external distribution)
- filing and records management which can be both digital and conventional.

Ideally document management should be OLE-aware so it can easily share documents with other OLE (object linking and embedding) aware software in Windows and it should be Open Doc aware to do the same thing in UNIX. The worldwide web has injected some new requirements: Autodesk has its DWF (Drawing Web Format, dubbed 'Dwarf') which allows the inclusion of hypertext links in the drawing. SoftSource has its SVF (Simple Vector Format) as well.

Activity database (ADB)

Another useful application of databases is the Department of Health's ADB (Activity Data Base).[2] Simply, ADB is a database of thousands of images of hospital equipment and furniture which correlate with their respective activities and processes, e.g. 'clinical handwashing' activity automatically produces a suite with a basin, wrist-action taps, soap dispenser, disposal bin, etc. (Figs 21 and 22).

This computerized information system, which is revised and updated periodically, is designed to help project and design team members by providing comprehensive design briefing material for individual spaces/rooms in hospitals and other health buildings and by defining users' needs more precisely. It is based on

- activity spaces rather than rooms, and thus includes unenclosed areas such as waiting bays/recesses, corridors and stations, and also allows more than one set of activities in a particular space

Fig. 21. Activity database — consulting/examination and two-sided couch access: (a) room data sheet; (b) room environmental data; (c) room design character; (d) schedule of components by room (courtesy of NHS Estates)

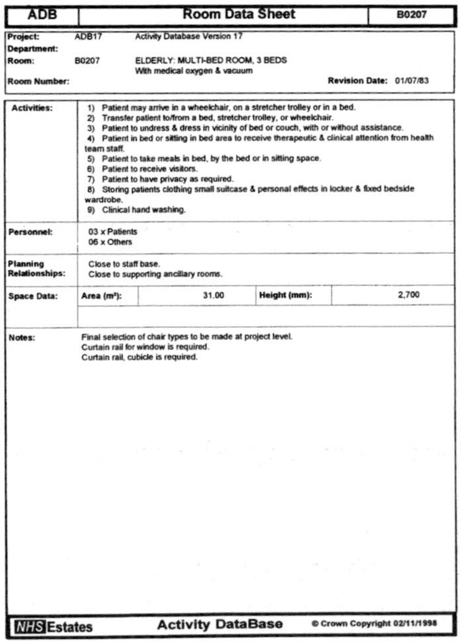

ADB | Room Data Sheet | B0207

Project: ADB17 Activity Database Version 17
Department:
Room: B0207 ELDERLY: MULTI-BED ROOM, 3 BEDS
With medical oxygen & vacuum
Room Number: Revision Date: 01/07/83

Activities:
1) Patient may arrive in a wheelchair, on a stretcher trolley or in a bed.
2) Transfer patient to/from a bed, stretcher trolley, or wheelchair.
3) Patient to undress & dress in vicinity of bed or couch, with or without assistance.
4) Patient in bed or sitting in bed area to receive therapeutic & clinical attention from health team staff.
5) Patient to take meals in bed, by the bed or in sitting space.
6) Patient to receive visitors.
7) Patient to have privacy as required.
8) Storing patients clothing small suitcase & personal effects in locker & fixed bedside wardrobe.
9) Clinical hand washing.

Personnel: 03 x Patients
06 x Others

Planning Relationships: Close to staff base.
Close to supporting ancillary rooms.

Space Data: Area (m²): 31.00 Height (mm): 2,700

Notes: Final selection of chair types to be made at project level.
Curtain rail for window is required.
Curtain rail, cubicle is required.

NHS Estates Activity DataBase © Crown Copyright 02/11/1998

(a)

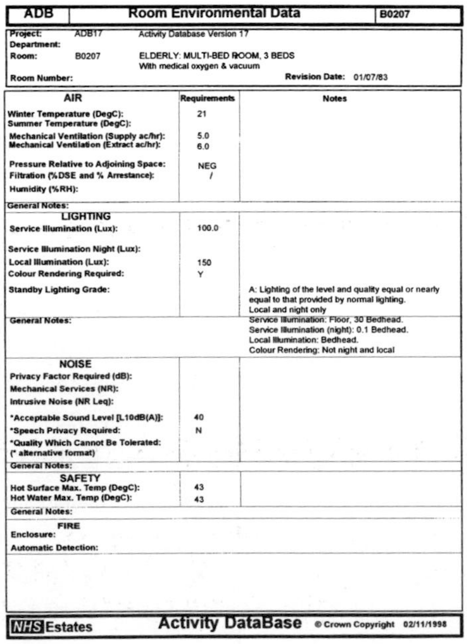

ADB | Room Environmental Data | B0207

Project: ADB17 Activity Database Version 17
Department:
Room: B0207 ELDERLY: MULTI-BED ROOM, 3 BEDS
With medical oxygen & vacuum
Room Number: Revision Date: 01/07/83

AIR	Requirements	Notes
Winter Temperature (DegC):	21	
Summer Temperature (DegC):		
Mechanical Ventilation (Supply ac/hr):	5.0	
Mechanical Ventilation (Extract ac/hr):	6.0	
Pressure Relative to Adjoining Space:	NEG	
Filtration (%DSE and % Arrestance):	/	
Humidity (%RH):		
General Notes:		
LIGHTING		
Service Illumination (Lux):	100.0	
Service Illumination Night (Lux):		
Local Illumination (Lux):	150	
Colour Rendering Required:	Y	
Standby Lighting Grade:		A: Lighting of the level and quality equal or nearly equal to that provided by normal lighting. Local and night only
General Notes:		Service Illumination: Floor, 30 Bedhead. Service Illumination (night): 0.1 Bedhead. Local Illumination: Bedhead. Colour Rendering: Not night and local
NOISE		
Privacy Factor Required (dB):		
Mechanical Services (NR):		
Intrusive Noise (NR Leq):		
*Acceptable Sound Level [L10dB(A)]:	40	
*Speech Privacy Required:	N	
Quality Which Cannot Be Tolerated: (alternative format)		
General Notes:		
SAFETY		
Hot Surface Max. Temp (DegC):	43	
Hot Water Max. Temp (DegC):	43	
General Notes:		
FIRE		
Enclosure:		
Automatic Detection:		

NHS Estates Activity DataBase © Crown Copyright 02/11/1998

(b)

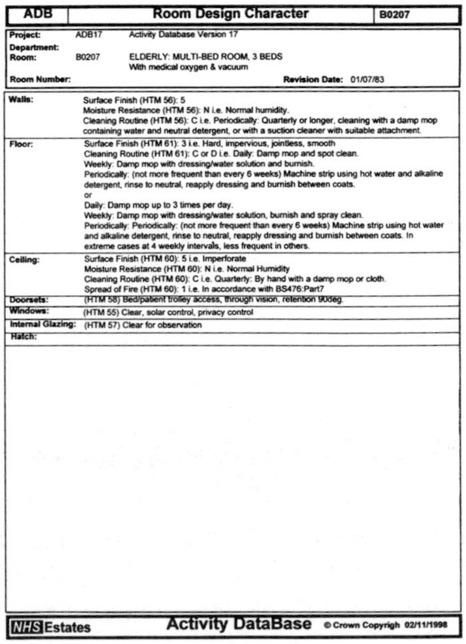

ADB | Room Design Character | B0207

Project: ADB17 Activity Database Version 17
Department:
Room: B0207 ELDERLY: MULTI-BED ROOM, 3 BEDS
With medical oxygen & vacuum
Room Number: Revision Date: 01/07/83

Walls:	Surface Finish (HTM 56): 5 Moisture Resistance (HTM 56): N i.e. Normal humidity. Cleaning Routine (HTM 56): C i.e. Periodically: Quarterly or longer, cleaning with a damp mop containing water and neutral detergent, or with a suction cleaner with suitable attachment.
Floor:	Surface Finish (HTM 61): 3 i.e. Hard, impervious, jointless, smooth Cleaning Routine (HTM 61): C or D i.e. Daily: Damp mop and spot clean. Weekly: Damp mop with dressing/water solution and burnish. Periodically: (not more frequent than every 6 weeks) Machine strip using hot water and alkaline detergent, rinse to neutral, reapply dressing and burnish between coats. or Daily: Damp mop up to 3 times per day. Weekly: Damp mop with dressing/water solution, burnish and spray clean. Periodically: Periodically: (not more frequent than every 6 weeks) Machine strip using hot water and alkaline detergent, rinse to neutral, reapply dressing and burnish between coats. In extreme cases at 4 weekly intervals, less frequent in others.
Ceiling:	Surface Finish (HTM 60): 5 i.e. Imperforate Moisture Resistance (HTM 60): N i.e. Normal Humidity Cleaning Routine (HTM 60): C i.e. Quarterly: By hand with a damp mop or cloth. Spread of Fire (HTM 60): 1 i.e. In accordance with BS476:Part7
Doorsets:	(HTM 58) Bed/patient trolley access, through vision, retention 90deg.
Windows:	(HTM 55) Clear, solar control, privacy control
Internal Glazing:	(HTM 57) Clear for observation
Hatch:	

NHS Estates Activity DataBase © Crown Copyright 02/11/1998

(c)

- the number of people who use the spaces regularly or intermittently (e.g. patients, others, etc.)

- the relationship of the space/room to other spaces/rooms (e.g. close to staff base, ward activity to be visible from room, etc.)

- a list of equipment (e.g. chair, table, etc.)

- environmental and services engineering requirements (e.g. air temperature 18°C; lighting, 150 lux local illumination; noise, 40L10dB (A) acceptable sound level, etc.).

Output from the ADB consists of three kinds of information.

(i)   *Activity space data sheets* — commonly referred to as A-sheets which provide design briefing information about the range of activities that normally take place within a particular space (e.g. (a) patient may arrive on foot; (b) patient may arrive in a wheel chair, on a stretcher trolley or in a bed . . . (f) patient in bed, sitting by bed or in sitting

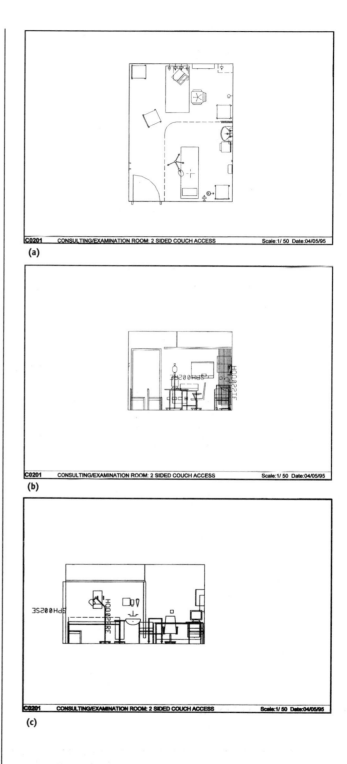

Fig. 22. Activity database — consulting/examination and two-sided couch access: (a) plan; (b) section elevation BB; (c) section elevation AA (courtesy of NHS Estates)

(d)

space to read, write, listen to radio, view TV and use external telephone . . . (k) clinical hand washing . . . (m) use of mobile X-ray machine . . . (n) use of monitoring/diagnostic equipment, etc.); the general requirements for the proper functioning of that particular space (e.g. lamp indicating call system (other than patient/staff), wall mounted; socket outlet switched 13 amp single, ac, wall mounted; rail curtain, glazed screen/door vision panel, etc.).

(*ii*) *Activity unit data sheets* — commonly referred to as B-sheets which record the functional requirements of a particular activity (e.g. cleansing: clinical handwashing; bed/cot care: piped medical oxygen vacuum; chair: upright, stacking, etc.). An activity unit comprises a recommended arrangement of equipment, furniture and engineering terminals appropriate to a particular activity. Each item in an activity unit is assigned to one of the equipment groups (group 1 — items supplied and fixed within the terms of the building contract, group 2 — items which have space and/or building construction and/or engineering service requirements, and are fixed within the terms of the building contract but supplied under arrangements separate from the building contract; group 3 — as group 2 but supplied under arrangements separate from the building contract, possibly with storage implications but otherwise having no effect on space or engineering service

requirements, and group 4 — items supplied under arrangements separate from the building contract, possibly with storage implications but otherwise having no effect on space or engineering service requirements).

(*iii*) *Activity space components summary sheets* — commonly referred to as C-sheets which record the total quantities of components in groups 1, 2, and 3 extracted from all the B-sheets that have been listed on individual A-sheets.

The Activity data bank was originally conceived as a means of ensuring that the design and the equipping of a given room or space proceeded from the knowledge of the activities that would take place within it, and an understanding of the ergonomic consequences of these activities.[3] The data bank suffers from a lack of development.

References

1. Miles D. Electronic systems can provide a foolproof method for the storage and retrieval of drawings and documents. *Architecture Today,* 21, Sept. 1991, 75.

2. Department of Health Building Directorate Branch HBD3. *Guide to 'A and B' activity data sheets and their use in health building schemes.* London, 1980.

3. Millard G. *Commissioning hospital buildings: a King's Fund guide.* King Edward's Hospital Fund for London, 1981, 36–38.

Data manipulation and mathematical analysis

Introduction

This chapter is concerned with scientific computing and in particular data manipulation and mathematical analysis including operational research simulation. The main drivers behind scientific computing are the problems that arise from science and engineering that need to be solved. Data manipulation can be defined as using a computer to process strings of data in order to provide answers according to some programming instructions. It is essential for scientific research/development (e.g. models for weather prediction and environmental conditions, high-energy physics, simulating experimental mathematics, etc.) and industrial applications (e.g. controlling manufacturing machinery, robots, assembly lines, inventory, phone switching networks, etc.). As most research is specialized and industrial tasks are either unique or so specific to each circumstance, the necessary applications and software tools are usually custom-built or proprietory, although basic applications can be purchased and used as building blocks for customized tools. Of importance to architects and engineers is an understanding of the main activities associated with organization, management and manipulation of data and mathematical analysis.

- *Data collection* involves the gathering of large volumes of data (both input and output) from furniture and equipment inventories, from book-keeping records, from invoices, time cards, purchase orders, stock movements, etc. as business transactions are occurring.

- *Data entry* involves coding and entering collected data into the computer for processing.

- *Data processing* involves a number of processes such as sorting, comparing, calculating, inserting, merging and presenting. During sorting, data files are held in a particular sequence, for example, a stock file may be kept in a number order, a wages file may be kept in employee number order, and the numbers are allocated in sequence as employees join the practice. To update the files transactions must be in the same order as the files and will need to be sorted appropriately. Comparing involves checking one data item against another to determine a course of action, for example whether a stock level is below reorder level. During calculations

the computer uses mathematical formula or functions as required by the program. Inserting involves entering new customers or employees into the records so that they appear in the correct sequence. The merging process entails updating a file by changing the information in it using the data from the latest transactions and therefore combines comparing and calculating functions. The information that is the result of the processing must be presented in a form which facilitates understandings and analysis and, in most cases, may be used to derive working principles.

Data processing (DP)

Data processing refers to the use of the computer to record, store, retrieve, analyse and communicate data. In particular, this covers the processing of business transactions such as producing the monthly payroll for an organization, or sales and purchases, although it also includes scientific and engineering 'number crunching' applications. It is distinguished from either word or image processing, or speech or music processing and allows the same advantages as other kinds of computer applications, i.e. it is fast, labour-saving, accurate (minimizing error) and cheap. Disadvantages of computer data processing include maintenance costs, inflexibility of procedures compared to manual methods, equipment breakdowns, which can be catastrophic, and concerns about computer security. Data processing activities include transaction processing, reporting, dealing with inquiries and maintaining files so that reference data are up-to-date.

Three major modes of processing data can be identified

(*a*) batch data processing

(*b*) on-line and real-time data processing

(*c*) distributed data processing.

In batch processing, transactions are grouped together in batches and processed as a group/batch. Most activities are not time-critical as long as the results are produced within an agreed or specified period of time — for example once every few hours, once a day, weekly or monthly (e.g. payrolls) and so on. These examples typify activities where instantaneous data processing is either unnecessary or would be too expensive. On-line and real-time processing is carried out if information is needed which is current. The mode of processing is then called 'real-time' (data is entered into a computer, processed, and results returned sufficiently quickly to affect the functioning of the system at that point in time) — for example booking systems (airlines, theatres, hotels, etc.).

Distributed processing is carried out in those situations where the processing of data is decentralized throughout an organization, with at least two geographically separated processors linked together. Data processing can be centralized, in a central data processing department, or decentralized in a number of user departments, where data entry is controlled and processed via on-line terminals. In bureau processing, a client firm sends its data for processing to a bureau which provides hardware and software (compare with outsourcing).

Statistical data processing

Statistics is the collection, classification and analysis of data. This is an important area for both architects and engineers. While most word processing packages offer similar basic features, the range of available statistics software is far broader. It is therefore far more important to know exactly what you require from your statistical software in order to be certain of choosing the most suitable package.

The first question to ask when choosing a statistics software package concerns the type of data likely to be available? Does it involve patient records, pharmaceutical treatments, a customer database or sales records, for example? Thereafter a host of other factors need to be considered

- are you interested in predicting future events or analysing past events?
- will you be analysing data by categories or discerning factors?
- how much data do you have, and in what form is it held?
- do you require SQL (Standard Query Language) connectivity?

Because spreadsheet packages (such as Excel, Lotus 1-2-3 and Quattro Pro) offer a range of basic statistical functions, the role of the specialist statistics packages has become unclear. If this specialist software is a tool, then what is it supposed to be used for, and who is supposed to use it? Further considerations in your choice of package are the output quality required, whether analyses will be graphical or numerical and the user's likely level of expertise. Finally, you must consider carefully whether your need for statistical data processing is likely to grow and if you want a general-purpose package with many procedures and routines.

Specialist Statistics Software includes

- S-Plus for Windows 3.2
- SPSS for Windows 6.1 (SPSS is the world's leading supplier of statistical data analysis software and services)
- Stratgraphs Plus for Windows 1.1
- Statistica for Windows 4.5
- Arcus Pro Stat 3.23
- Microfit 386 3.21
- Unistat for Windows 4.0.

Most of these packages contain a range of procedures and routines. Examples include distribution functions, parametric and non-parametric tests, cross tabs, chi-square, correlation analysis, survival analysis (Kaplan–Meier, Simple life Peto's Log-Rank, Wilcoxon test, Wei–Lachin analysis), analysis of variance, regression, categorical analysis, ANOVA, cluster analysis, facilities for handling complex numbers, vectors, matrices and other aspects of real mathematical computing. Potential users should think about the type of analyses required, and select the package(s) which are likely to be most suitable.

Simulation studies/computer modelling and IT, and prediction of building performance

Mathematical techniques to evaluate solutions to design problems such as finite-element analysis and mechanism analysis have been common since the 1960s. The continuing development of easy to use interfaces has gone hand-in-hand with the ability to handle ever more complex problems. Also, the number of manufacturers, engineers and designers with calculation software has been growing. Computer simulation models are now available to deal with lighting, thermal response ventilation, plant performance and so on. For example, the computer simulations for Airflow Analysis involved CFD (Computational Fluid Dynamics), for both data input and output. CFD technology can be used to integrate detailed air flow analysis and to predict the impact of natural ventilation on cooling loads and internal comfort. The technology can be used to analyse the performance of air purging systems in critical spaces such as laboratories, allowing the output visualization to be run in real time and presented as a video.

Programs such as Phoenix-VR provide a 'VR' (virtual reality) interface which allows models to be constructed in 3D, with components such as doors, windows and furniture added from a standard library of clip-art for lighting design. The simulation of true physical characteristics of light and materials allows interactive exploration and analysis, making design information more meaningful. According to Concord Sylvania, computers do make light work and increasingly they come into their own for lighting designers, both for evaluation and presentation.[1,2] This is shown by their computer system which uses five programs to evaluate luminaire data and to apply it to interior and exterior lighting products and projects, including the following.

- The evaluation of luminaire data from photometric information. The data is stored and can be presented in formats appropriate to different national formats and requirements.

- The second and third programs are lighting programs for regular or irregular interior spaces. The computer facilities can encompass any shape of interior space — sloping walls, barrel vaults, raked floors — and quickly and precisely plot luminaire positions. 3D grids of illuminance are produced as well as predictions of illuminance in true perspective from any viewing angle. Colours can be preset, so allowing for colour prediction that will, for example, show the pink tone that reflects on to a white wall from a red carpet. Doors, windows and pictures can be incorporated into the evaluation of reflections, not only improving the accuracy of the scheme, but giving the client a better recognition factor when printed data includes characteristics of a particular space.

- The exterior lighting program includes still more data to allow for aiming angles. Floodlighting can be calculated precisely and there are layouts for sports stadia and tennis courts where even the shadows and TV camera positions can be plotted.

- The daylight program is particularly useful when dealing with atria

or the large expanses of glass in modern buildings. The intensity of daylight can be balanced with artificial light sources at the exact point of the interior where extra light is needed.

Typically, another of Concord's programs, SLI-WIN (Windows-based Professional Lighting Design Program), includes a series of templates showing examples of different rooms and applied lighting calculations as well as colour pictures and an explanation of the design rationale. This feature is designed to help the occasional user, like the electrical contractor, who increasingly finds himself confronted by greater demand on his/her skills and needs to be able to prepare a lighting scheme to supply a calculation and print-out of illuminance requirements. The user enters the room plan (rectangular or L-shaped) and ceiling height, chooses a lighting position from the range of lamps and fittings available and specifies the reflectances of the wall and floor-covering materials. SLI-WIN shows the selected points of the lighting levels. Alternatively, the user can enter the desired illuminance level at the working plane and the program will calculate the number of fittings required and their positions. As the program calculates the iterations of reflected and absorbed light as well as direct light from the light source it is extremely accurate.

Other proprietary programs such as Radiance (written by Greg Ward of Lawrence Berkeley Laboratory, USA, first released in 1989 and available on the Internet) can be used to trace light from natural and artificial sources and to generate a full 3D image of an illuminated space — 'photo-accurate visualization'. Successful application of Radiance software includes the lighting bridge project by the French lighting company Aldex and by the Ecadap Centre at de Montefort University, Leicester, for a Peter Foggo Associates scheme for Stanhope Properties.[3]

It is becoming increasingly common for computer modelling to be used both to predict the performance of buildings and to exchange technical data between building professionals, providing new opportunities to improve design and construction processes. Performance models for increased efficiency cover the following.

- Acoustics, e.g. programs for testing particular spaces (speech rooms, auditoriums, etc.) including effectiveness of fire alarm systems.

- Condensation, fire risk assessment, health and safety appraisal, e.g. the evaluation of potential risks using particular materials in certain constructions, examining scenarios to determine effectiveness of escape routes.

- Energy, e.g. models to determine consumption, evaluating running costs, energy management benchmarking software such as that for higher education institutions which allows calculation of individual site performance benchmarks for comparison against specific benchmarks and for identifying areas where action is required owing to poor performance.

- Heating and ventilation, e.g. calculation of plant heating capacity, cooling loads on air conditioning plant, air conditioning models and sizing of service ducting and pipes, programs for assessing air flows, etc.

- Lighting, thermal analysis and performance, e.g. calculation of daylight, artificial lighting levels.

- Materials and structures, e.g. analysis of the behaviour of specific materials and structures.

- Civil and structural engineering, e.g. translating expressive and abstract sculptural shapes with complex geometry into real physical structures, or terrain-modelling or static and dynamic analysis of 2D and 3D structures.

- Geotechnical appraisal; or the creation of a digital ground model involving the conversion of land survey data or sub-surface data into a mathematical representation of a site, from which calculations and drawings can be extracted and road and drainage design produced; or solving retaining walls, piling, slope stability and ground movement problems.

- Design and detailing of foundations, superstructure, roads and car parking.

The OasysBEANS system (Building Environmental Analysis System) available from Oasys, London, is an example of some of the comprehensive suite of programs available for appraising performance. The interactive suite of programs covers fabric analysis (FABRIC), window solar gains and shading (WINDOW), steady state heat loads (HEAT), thermal analysis (THERM), lighting analysis (LIGHT), room analysis (ROOM), Building Energy Code Part 2 analysis (BENC2) and ventilation analysis (VENT).

Lack of agreed standards regarding data sharing is one of the problem areas to be resolved.

Energy consumption

The total fuel used in a typical house can be estimated using computer programs that incorporate the Building Research Establishment Document Energy Model (BREDEM). BREDEM is one of the methods of showing that the design conforms with Part L of the Building Regulations. It was developed using experience gained from measurements made on a large number of occupied dwellings as well as research into many aspects of their design.

It underpins the given *Standard Assessment Procedure* (SAP) for home energy rating (a scale of 1–100 according to energy efficiency, taking into account space and water heating costs). It is also used by the two commercial labelling services from the National Energy Foundation and MVM Starpoint System in order to estimate annual fuel consumption for space and water heating, expressed on a scale of 1 to 5 stars for the most energy-efficient housing. Energy ratings in themselves offer a simple method of informing householders about the overall efficiency of a home — a low SAP value means that the house will be expensive to heat and will therefore have high bills. A new build specification standard, for example 80, 90, or 100 SAP, is a likely requirement for new projects.

Intelligent buildings

The term intelligent building originated in North America in 1980, stimulated by deregulation of telecommunications. By the mid 1980s, there was a realization that thoughtful design, specification and building management could help an organization absorb IT and improve its competitive edge. Increasingly in the 1990s, the effective application of IT is recognized as an essential resource for a successful business.

The main goals of the research project *The intelligent building in Europe*, undertaken by DEGW and Technibank together with their European group-related subsidiaries, defined a new European user-focused model of building intelligence and produced a competitive analysis and forecast for the European intelligent building marketplace. Three goals were defined for an intelligent building,[4] one which 'provides a responsive, effective and supportive intelligent environment within which the organization can achieve its objectives'. These are

- building management — the management of the building's physical environment using both human systems (facilities management) and computer systems (the building automation system)

- space management — the management of the building's internal spaces over time: the overall goals of effective space management are the management of change and the minimization of operating costs and the satisfaction of the end user

- business management — the management of the organization's core business activities and working patterns. In most cases this is characterized as a combination of ideas and information in many forms.

Successful intelligent building solutions depend on three layers.

- Providing effective building shells which can absorb IT and allow the organization to grow and change.

- Applications of IT (as defined by building, space and business management systems) to reduce costs and improve performance.

- Provision of integrating technologies and services.

The study also identified the trend of converging technologies within intelligent buildings depicting this in the three stage 'IBE Integration Pyramid' (Fig. 23).[5] Stage 1 is Building Automation (BA), where security and service control systems are integrated. Stage 2 is Integrated Communication (IC), where office automation and business management systems are integrated. The computer integrated building (CIB), where all systems are interconnected, is the final stage — the ultimate result of increasing integration for building control and IT systems into a unified whole.

Current intelligent building systems feature integrated energy management and security systems based around electronic sensors and digital control panels. The microprocessors in these systems perform sophisticated analyses of incoming signals and beam the information on the PC display screens to assist the building manager. User feedback from sensors helps to use the buildings more effectively. For example, energy consumption levels can be displayed so that if a set level is exceeded, appropriate action can be taken. To date there have been problems associated with these systems and evident since the 1970s with the impact of IT on building services — for example, overheating in computer-intensive offices, ceilings collapsing under the weight of copper cabling, over-filled raised floors and ceiling tiles which do not lie flat.[6] Another problem with these integrated systems is that even when they work well they are often difficult to maintain, such that the potential benefits are over-

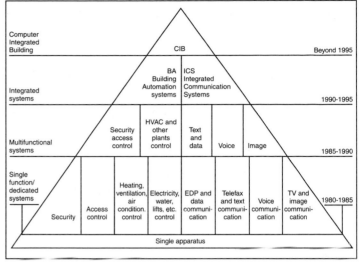

Fig. 23. IBE Integration Pyramid

shadowed by increased costs. Further problems that concern the IT industry are, for example, bandwidth limitations, security, legal issues and immature technologies.

Increasing integration of new and more accurate computer technological programs allows greater consideration of environmental and management issues such as life-cycle costs, improved efficiency and effectiveness, green design, decision-making concerning the most appropriate building fabric and building service configuration, and addressing design and management problems, e.g. inflexible and incomprehensible energy management systems, poorly shaded windows, etc.

References

1. White, S. Computers make light work. Energy in *Buildings & Industry*, June 1996, 14.

2. Gosney, J. Reflecting the right image. *Light,* January 1996, 26-31.

3. *ibid.*

4. Bernard William Associates. *Facilities Economics incorporating Premises Audits.* Building Economics Bureau Ltd, Kent, 1994, 7-37–7-39.

5. DEGW/Technibank. *The Intelligent Building in Europe (IBE).* 1992.

6. Southwood, B.W. Can you dig IT? The information challenge: Future Trends Information Technology. *Building Services Journal,* 1997, 26.

CAD/CAM/CAE and multi-media

Introduction

This chapter is concerned with CAD (computer-aided design), CAM (computer-aided manufacture) and CAE (computer-aided engineering). It starts by considering CAD and then goes on to look at its historical background, including specific features such as computer graphics and CAD application areas. Its relationship to CAM and CAE is also discussed. This includes looking at how traditionally CAD software development has mimicked the hardware tools (pens, paper, paint brushes, etc.) used in practice. In general most CAD systems have to earn their keep in two main ways, either as drafting tools for complex drawings/general arrangements or as design tools and presentation devices for 3D-modelling and rendering.

CAD (computer-aided design): definitions

CAD is defined as the use of a computer program to generate designs, normally in the form of dimensional drawings. CAD uses the Cartesian co-ordinate system (X, Y, Z), defined by René Descartes, the 17th century French mathematician and philosopher, to allow representation of objects in the 2D and the 3D world. Therefore the system allows lines, arcs and circles to be drawn as geometric 'entities' on the computer screen which can then be scaled to output the appropriate drawing using a pen plotter or similar device.

The CAD system builds up a database of the various entities created, and can then store them on computer disk for future use or modification. There are many benefits of using CAD systems — for example design changes can be made easily, and because the design is stored electronically there are none of the problems associated with physical storage. Also buildings are rarely modelled at full scale so that the use of computer models offers great advantages over scale cardboard models — a tool for redesign, for checking tactile, ergonomic and functional aspects and a tool to produce drawings quickly and accurately. But how did it all begin?

CAD: a brief history

The history of CAD is, like the rest of the computer industry, deeply entwined with developments in the US military and space industry starting in the late 1940s. The first interactive CAD package, the SAGE (Semi-Automatic Ground Environment) air defence system was developed in 1947 to monitor all the radar stations in the US as part of the defence strategy against the perceived 'Soviet threat'. The SAGE system was the first to use command and control (CRT — cathode-ray tube) consoles on which operators identified targets by pointing at them with light pens.

By 1951, the US space effort was well under way and Massachusetts' Institute of Technology (MIT) Lincoln Laboratory was developing the world's first super-computers, called the TX series. During the period 1962–3 a thesis entitled *Sketchpad; a man–machine graphical communication system* by a student at MIT (Ivan Sutherland), described the first truly interactive computer graphic system, and in so doing not only influenced all major developments in CAD, computer graphics and virtual reality but enabled Sutherland to be known as the father of computer graphics.[1] Detail of the displayed image of the SKETCHPAD II system by Ivan Sutherland on the TX-2 computer of Lincoln Laboratory, MIT, is shown in Fig. 24; a perspective view and three orthographic projections of a chair.

Although limited to drawing in 2D, the early version of the SKETCHPAD system showed that the designer could, for the first time, interact with the computer graphically via the medium of display screen and light-pen. The advent of graphic devices was helpful in the growing computer applications in building design.[2]

Over the next ten years, corporations such as IBM, Lockheed Aircraft, McDonnell Douglas and Xerox took CAD even further, so that by the early 1960s the first vector-related 2D/3D packages were developed for mainframe use. Events at GM (General Motors) and DEC (Digital Equipment Corporation) are particularly significant. At General Motors, the DAC-1 console was installed by IBM on a 7094 computer. The DAC-1 was used to test human–machine interaction for the first time and showed one-third increase in productivity. Digital Equipment Corporation also produced its first computer, the PDP-1, which was followed by PDP-9 in 1966 and then PDP-11 (which opened the door to

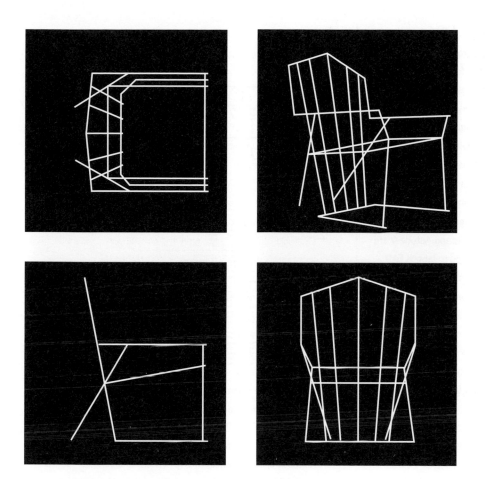

Fig. 24. Detail of displayed chair image from Ivan Sutherland's SKETCHPAD II system (courtesy of Taylor and Francis Ltd)

successful Macintosh 128K computer together with a free software package, MacDraw, an object-oriented, vector-based, 2D drawing package. By this time Apple had gained a reputation for innovation, fuelled by the mythology of its small beginnings in a garage, which had caught the public imagination. Suddenly CAD was available on a reasonably priced, easy to use microcomputer, enabling architectural schools and small practices across the US and the UK to acquire Macintosh computers and to start producing respectable and competitive imagery, using simple output devices such as the Imagewriter. The newly established Claris then produced ClarisCAD, which led to the Macintosh being known as a user-friendly and creative tool, and inspired new software development.

By late 1986, the great variety of software that was available for 2D drawing, 3D visualization and image retouching raised many questions about the future of the Macintosh as a design tool. The need for colour images, greater memory and faster machines to handle complex CAD graphics quickly and efficiently, and for large-scale work (rather than crude 'wireframe graphics'), all had to be addressed. The fast Macintosh II with 8Mb of RAM and colour was released and immediately unleashed the 3D potential for visualization requirements, such as information storage for models, perspective generation, surface calculations, rendering with variable lighting and animated colour 'walk-arounds' or 'fly-throughs'.

As a result, by early 1988 large software corporations that had previously specialized in engineering and TV animation software on other platforms became interested. In particular two software corporations in the US — Autodesk® and Intergraph® (Bentley Systems) — created their popular CAD packages AutoCAD and MicroStation. MicroStation has its origins in the early 1970s as the product IGDS by

CAD for DEC), which in turn was succeeded by PDP-15.

An important predecessor of the PC was the IBM 360, introduced in 1964 helped by Lockheed Corporation along with General Motors and McDonnell Douglas. This was followed by the 370, 30XX and 4300 series over the next 20 years of progress of IBM in CAD computer support. During the 1970s IBM developed AutoCAD, the CAD package which set the standard by which CAD was to be judged. This package was virtually targeted at large architectural practices but was, by the early 1980s, extended to CAD on the desktop. Until then CAD was mainly characterized as used by technicians digitizing plans for output to A1/A0 plotters; an image which not only dominated the common perception of CAD users (designers and architects) but also limited development.

However, in 1984 Apple Computers released the immensely

Intergraph Corporation on the UNIX platform and used sophisticated tools to relieve the burden of producing manual production drawings as well as containing advanced visualization tools for creating computer-generated models. The enormous costs associated with the purchase of a UNIX system meant that only the very large design practices could afford the CAD application. The migration to PC and Macintosh under the name MicroStation brought with it full design integration, powerful production drafting and visualization tools based on almost 20 years of development on the VAX and UNIX platforms including a lot of 'big system capabilities', for example networking, distributive databases, supporting concurrent designers and consultants through reference files. Bentley Systems and Autodesk are clear market leaders. Autodesk was the first company to launch CAD on the PC platform and has been the world leader since, with over a million users worldwide of its acclaimed AutoCAD product. Comments have been made that AutoCAD was still a drafting production tool, dependent on technological advances made to PCs and Intel processors; this is despite the fact that a decade or so after its introduction it was established as a standard application for CAD in the design community and universities. Upgrades only included minor fixes and additional features (e.g. 3D or surface modelling, photorealistic renderings, animation, etc. to be bought separately) that did not take the next step from a drafting tool to a total integrated design application, but were a series of complementary products to be used in conjunction with AutoCAD standard application. The diversity of the design profession did not seem to help AutoCAD's development.

Others are CADlogic®, Leonardo Computer Systems®, CODEC® and so on. The architectural CAD markets of the US and Japan matured first, before the UK, and were followed by the rest of western and then eastern Europe. SpiritCAD® Software (based on the American system, DataCAD®, with its logical architectural approach) dominates the German architectural market (marketed by Soft-Tech as Spirit/DataCAD).

Of importance in looking at the development of CAD systems in the UK is the influence of the public sector authorities such as the Property Services Agency (PSA), Department of Health and Social Security (DHSS) with the HARNESS system, Department of the Environment with the CEDAR2 system, West Sussex County Council with the SCOLA-based system, Clwyd County Council and various universities — notably the University of Strathclyde (SPACES, GOAL, PARTIAL, PHASE, AIR-Q by ABACUS), the University of Edinburgh, the University

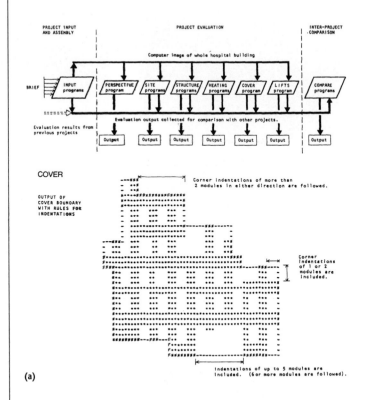

(a)

of Sheffield (GABLE system) and the University of Liverpool (CARBS).

West Sussex County Council's design offices in Chichester led the pioneering work on the development of the interactive graphics computer-aided building system (Ray Jones, 1968, Paterson 1972, 1974).[3–5] The system for the SCOLA industrialized component building system used refreshed cathode-ray tube graphic terminals equipped with light pens and was driven by an in-house IBM System/370 computer to produce descriptions of buildings, cost analyses of designs, environmental and structural evaluations and automatically-generated construction documentation for a serial tendering project programme. The system was discontinued in 1974 following a change in policy.

Another UK development is HARNESS (Hoskins and Bott 1972, Meager 1972, 1973).[6–8] Sponsored by the DHSS, the design system represents a hospital assembled by arranging standardized, pre-designed hospital departments along a circulation spine. The high level of standardization allowed complete automation of the design process with the computer system performing structural, environmental and cost evaluation, automated generation of layouts, perspectives

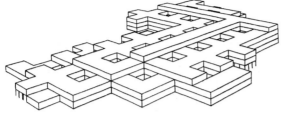

HIGH-LEVEL PERSPECTIVE FROM ONE OF FOUR POSSIBLE VIEWPOINTS

BUILDING DATUM
SOIL TYPE 1
SOIL TYPE 2
SOIL TYPE 3

FILL CUT FILL CUT

SECTIONAL VIEW SHOWING
BUILDING ON LEVELLED SITE

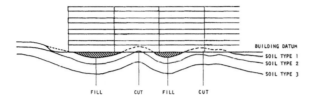

LIFTS

```
*CUT 1
CUBIC METRES

  LOOSE   COMPACT   FIRM   ROCK   FILL
  1659     5335    1573     0    13414
   565      303      98      0     3159

        SITE ANALYSIS
```

OUTPUT FROM CUT AND FILL ANALYSIS

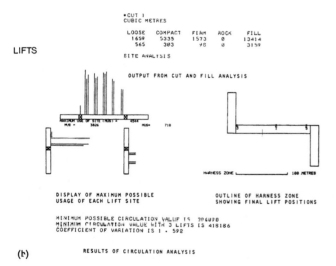

MAXIMUM USE OF SITE (MUS) =
MUS = 3826   4544   MUS= 710

HARNESS ZONE   100 METRES

DISPLAY OF MAXIMUM POSSIBLE
USAGE OF EACH LIFT SITE

OUTLINE OF HARNESS ZONE
SHOWING FINAL LIFT POSITIONS

MINIMUM POSSIBLE CIRCULATION VALUE IS 306090
MINIMUM CIRCULATION VALUE WITH 3 LIFTS IS 418186
COEFFICIENT OF VARIATION IS 1 . 592

(b)   RESULTS OF CIRCULATION ANALYSIS

HEATING

```
RATE OF HEAT LOSS IN WATTS FOR UNIT TEMPERATURE DIFFERENCE

DEPT NAME  CONDUCTION LOSS  VENTILATION LOSS  INTERNAL GAIN

MORTUARY     +1574.18640      +481.83984      +.00000000
STFF CHG     +2857.91760     +2658.11012      +.00000000
Y CNR W      +4635.11970     +2891.43984      +.00000000
PATH 1       +3092.19120     +1927.35936      +.00000000
PATH 2       +3464.39430     +1927.35936      +.00000000
ITU          +2141.78580     +2168.27928      +.00000000
A+E          +1458.05940     +2489.19920      +.00000000
OPD+PR       +7792.94790     +6625.29760      +.00000000
ENTRANCE     +2474.89700     +1445.51952      +.00000000
PHYS MED     +5366.49030     +6103.38464      +.00000000
G WARD       +9331.10280     +8673.11712      +.00000000
LINEN        +1000.66590      +240.91962      +.00000000
HSSU+*AN     +2621.94570     +1445.51952      +.00000000
CNPL+LIB      +817.66260      +481.83984      +.00000000
PAT SERV     +2255.93010     +1445.51952      +.00000000
PHARMACY     +1438.27650      +963.67968      +.00000000
X-RAY        +1924.90830     +2489.19920      +.00000000
ID UNIT      +2474.74440     +2891.03984      +.00000000
A WARD 2     +3649.60400     +8673.11712      +.00000000
DAY WARD     +1050.92640     +2168.27928      +.00000000
THEATRES     +4896.03160     +9636.79680      +.00000000
MAT WARD     +4321.20960     +8673.11712      +.00000000
A WARD 1     +4857.05510     +8673.11712      +.00000000
CATERING     +3473.75790     +3372.87808      +.00000000
MANAGMT      +3515.09410     +2489.19920      +.00000000
AMC          +1411.14900      +963.47968      +.00000000
MAT+SCBU     +7373.42190     +7958.35536      +.00000000
A WARD 3     +4551.12270     +5059.31032      +.00000000
EDUCATN      +2814.06340     +1927.35936      +.00000000
TRAINING     +3919.34790     +1927.35936      +.00000000
PAED W       +8097.58620     +8673.11712      +.00000000
STORES       +8899.38350     +2168.27928      +.00000000

HARNESS HEAT LOSSES

LEVEL 5       +8294.08410     +5942.69136      +.00000000
LEVEL 4 *     +3766.92120     +5942.69136      +.00000000
LEVEL 3      +10538.45640     +7923.56848      +.00000000
LEVEL 2       +3482.42930     +1980.49712      +.00000000

FUEL CONSUMPTION IN KWATT.HOURS IS +7509374.54
```

OUTPUT FROM HEATRUN

COMPARE

ELEMENT	NUMBER	UNITS	COST	% OF COST
EXCAVATION:				
LOOSE	1982	CU.M.	5946	0.3
COMPACT	9456	CU.M.	47280	2.5
FIRM	1748	CU.M.	6992	0.4
ROCK	0	CU.M.	0	0.0
FILL	15496	CU.M.	61984	3.3
BACKFILL	8042	CU.M.	16084	0.8
STRUCTURE:				
BEAMS	1713	NUMBER	301335	15.9
COLUMNS	3163	NUMBER	136032	7.2
FOUNDTNS	729	NUMBER	29700	1.6
SLABS	2181	NUMBER	471535	24.9
CLADDING:				
WALLS	1463	NUMBER	813428	42.9
EAVES	422	NUMBER	6752	0.4
			1897068	100.0

(c)

COST SCHEDULE OF STRUCTURAL COMPONENTS

and production documents. The aids of 'an integrated set of computer programs enabled a design team to input, evaluate and compare development plans' (Development Plan System). Figure 25 shows the synopsis of the Harness Hospital Computer-Aided Development Plan System. There were not many hospitals built using the HARNESS System, which was succeeded by the Nucleus Hospital System. Figure 26 depicts an AutoCad drawing of a Nucleus Adult Acute Ward (1990) for comparison.

The OXSYS system (Hoskins 1972)[9] was developed for the design of

Fig. 25. Harness Hospital Computer-Aided Development Plan System: (a) output of cover boundary with rules for indentations; (b) high-level perspective, output from cut and fill site analysis and results from circulation analysis; (c) output from heat run and cost schedule of structural components (courtesy of NHS Estates)

hospitals in the Oxford Method of Building[10,11] and was used for cost estimation from early sketch designs, structural and environmental analyses, semi-automatic design and for detailing and production of contract documentation. The fundamental concept was to replace the image of a building on paper with a readily accessible, easily-manipulated image on computer to allow analysis, evaluation and modification by every member of the team irrespective of their geographical location. User requirements were recorded as departmental and room data. Figure 27 shows examples of OXSYS outputs (i.e. room data, cladding schedules, layout and elevation).

The CEDAR 2 system (Chalmer 1972)[12] was developed by the Department of the Environment for Post Office buildings using the SEAC (South East Architects Collaboration) component system. The system had capabilities for cost estimation, daylight, thermal and acoustic analysis, detailed design framing and external walling, and for the production of documentation. CEDAR 3, a successor to CEDAR

2, was started in 1975 for use at the sketch design stage and in conjunction with the Property Services Agency's Method of Building (MOB) — a set of rules concerning dimensions, a set of ranges of preferred components, and a set of standard details rather than a component-based system. CEDAR 3 was aimed at comparing alternative building geometrics and site layouts with respect to capital and running costs. The system had facilities for building-description input editing, cost analysis, elevation selection, thermal analysis, daylight calculation and energy cost calculations.

Another development, CARBS (Computer-Aided Rationalized Building System) by the University of Liverpool Computer-Aided Design Centre in collaboration with Clwyd County Architects Department, Wales (Daniel 1973),[13] was basically for design evaluation and documentation.

The SPACES system by ABACUS at the University of Strathclyde in Glasgow was among the first attempts at an integrated computerized design system involving the computer in all stages (analysis, synthesis and appraisal) of the design process to develop a single package and SPACES was intended as a tool for the preliminary design of schools (Th'ng and Davies 1972).[14] The SPACES package represents a forerunner of today's commercially available CAD systems (Roberts 1990).[15] SPACES 1 covered the data analysis of spatial requirements for a proposed school — subject advisers listed for every stream in every year the number of pupils taking a subject together with the group age, the number of groups and the hours spent by the pupils (over one timetable cycle) in that particular subject. Output was tabulated as a schedule of accommodation. SPACES 2 represented each floor as a bubble proportional to the size of each area with the user able to move around these bubbles on the screen and convert them into rectangular spaces. The interaction matrix was used to define associativity between spaces. SPACES 3 was used to evaluate the layout produced in SPACES 2 by input of certain data relating to the geometry, activity within and construction of the building. CEDAR and HARNESS CAD packages in use at this time also followed the approach adopted by SPACES, i.e. the user working with a set building system.

Another development at ABACUS was GOAL (General Online Appraisal of Layouts) (Succock 1978),[16] an evaluative program designed along the lines of the SPACES 3 routine but able to work on a variety of building types. Other appraisal programs developed by ABACUS include PHASE (Package for Hospital Appraisal, Simulation and Evaluation) and the Air-Q program to aid the designer in

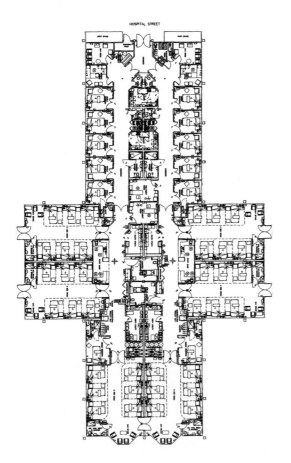

Fig. 26. AutoCad layout of a Nucleus Adult Acute Ward (courtesy of NHS Estates)

Room parts list

COMPONENT	COMPONENT DESCRIPTION	QUANTITY
5.31:1011	DAYLIGHT ESSENTIAL	1
5.31:1014	VENTILATION NATURAL ESSENTIAL	1
5.31:1025	INT DESIGN TEMP 20C	1
5.31:1031	AIR CHANGE RATE = 3	1
21.31:1010	WINDOW NORMAL CILL HEIGHT	1
21.31:1014	WINDOW CURTAIN TRACK	1
26.31:1011	INT DOOR VISTAMATIC PANEL	1
26.31:1012	INT DOOR PROTECTED	1
26.31:1015	INT DOOR HOLD OPEN	1
26.31:1018	INT DOOR HOSPITAL BED WIDTH	1
26.31:1027	INT DOOR SIGN - ROOM NO	
	+ DESIGNATION	1
27.31:1021	FINISHES GRADE B REGULAR WASHING	1
31.31:1010	BASIN WASH HAND NORMAL HEIGHT	1
31.31:1027	DOMESTIC HOT WATER POINT 40C MAX	1
32.6:1011	LOCKER BEDSIDE	5
32.8:1010	WARDROBE	5
32.12:1022	HOLDER PAPER SACK LARGE	2
32.12:1026	HOLDER THERMOMETER	5
32.12:1030	LIGHT EXAMINATION BEDHEAD	5
32.12:1038	MIRROR FULL LENGTH	1
32.18:1010	CURTAIN TRACK CUBICLE	1
32.18:1014	CURTAINS	1
33.1:1010	BED DIVAN	5
33.1:1018	CHAIR EASY	3
33.1:1022	CHAIR SEMI-EASY	2
33.1:1049	CHAIR STACKING	2
34.31:1021	PILLOW FOAM	5
34.31:1032	MATTRESS FOAM WITH COVER	5
40.31:1010	HEATING LTHW	1
50.31:1011	SOCKET 13 AMP TWIN	5
50.31:1012	SOCKET SHAVER	1
50.31:1014	TELEPHONE GPO JACK POINT	1
50.31:1016	NURSE CALL SYSTEM	1
50.31:1026	RADIO	1
60.31:1010	LIGHTING - FLUORESCENT	1
60.31:1016	LIGHTING - NIGHT	1

Cladding schedules

COMPONENT	ARB CODE	MANUFACT-CODE	COMPONENT DESCRIPTION	QUANTITY
21.1:22	S18	SP 18 30	SOLID H30 W6 30	1
21.1:23	S24	SP 24 30	SOLID H30 W6 30	2
21.1:24	S30	SP 30 30	SOLID H30 W6 30	11
21.1:220	S06	SP 06 24	SOLID H24 W6 30	9
21.1:221	S12	SP 12 24	SOLID H24 W6 30	20
21.1:223	S24	SP 24 24	SOLID H24 W6 30	8
21.1:224	S30	SP 30 24	SOLID H24 W6 30	20
21.1:230	H1183	u18 12 24 VS	HALFGLAZED H24 W12 C9	1
21.1:232	H1809L	09 18 24 VS FG	HALFGLAZED H24 W18 30 C9 LH	1
21.1:234	H3009L	09 30 24 VS FG	HALFGLAZED H24 W18 30 C9 LH	3
21.1:236	H1809R	09 18 24 FG VS	HALFGLAZED H24 W18 30 C9 RH	2
21.1:238	H3009R	09 30 24 FG VS	HALFGLAZED H24 W18 30 C9 RH	5
21.1:240	F1209	09 12 24 VS BG	FULL GLAZED H24 W12 C9	1
21.1:242	F1809L	09 18 24 VS FG BG	FULLGLAZED H24 W18 30 C9 LH	1
21.1:244	F3009L	09 30 24 VS FG BG	FULLGLAZED H24 W18 30 C9 LH	11

(a)

Fig. 27. Computer system for Oxford Method, OXSYS, typical outputs — (a) plan, room parts, lists, elevation drawing and cladding schedule; (b) 3D drawing (courtesy of Oxford Method of Building)

(b)

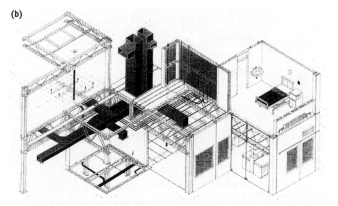

appraising the adequacy of any proposed passenger-handling system within an airport.[17]

The fairly general and comprehensive integrated system for housing design was developed initially by Ed CAAD for use by the Scottish Special Housing Association (Bijil 1971, 1972, *Applied Research of Cambridge*, 1975).[18,19] The system was used to generate working drawings (plans, elevations, sections, joists layouts, area and thermal analyses and plumbing isometrics).

Figure 28 shows one of these early CAD system developments, the system concepts of the Gable 4D series CAD system for Building Design comprising six modules: utilities; 2D drafting (IDS); 3D visualization (OMS); data base management (DMS); building modelling (BMS); and ground modelling (GMS). The Gable CAD system was developed at the University of Sheffield as a commercial system in the early 1980s.

- The 2D Integrated Drafting System (IDS) generated drawing files and allowed the user to define linestyles, hatching, area fill, text fonts, etc., with drawing files created by the utility plotter.

- The 3D Object Modelling System (OMS) enables the generation of 3D objects — perspectives, axonometrics, isometric and general parallel projects together with plans, elevations and sections and it allowed all views to be passed as drawing files to IDS.

- The Data Management System (DMS) was an interactive alphanumeric data base facility allowing user–defined data files which

checks, and with background interpretation systems automatically developing space and exterior models and passing to OMS. Plans, elevations, sections and perspectives could be sent to IDS.

The IDS, OMS and BMS files can be surveyed and schedules of quantities produced. The GABLE CAD system offered an integrated suite of systems all capable of being used separately or in any combination — exchanging information and together providing a comprehensive computer-aided design facility. Although envisaged as a 3D modelling system designed specifically with architects in mind GABLE had many similarities to AutoCAD and could not compete in the ready market of the workplace with such programmes which concentrated on the opposite end of the design spectrum — production drawings and visual-ization — and handled components (walls, floors, ceilings) rather than spaces. The production methods of buildings that emerged in the 1970s and 1980s such as the Oxford Method 3M, charac-terized by prefabricated trusses, rooflights, glazed cladding panels, partitions, bathroom and toilet 'pods', fixtures and fittings, etc., provided this ready market.

Software development in the UK also had a dedicated and high-profile following, notably Heron Associates using Modelshop® and MiniCad®, Mike Gold using Design Dimensions® and Leslie Fox Albin using ArchiCAD. In general there are now dozens of good CAD systems to choose from, some costing less than £100 with others costing many thousands of pounds. The difference is generally on perceived power, flexibility, customization and integration abilities. It is sometimes difficult to see exactly why some systems are so expensive when they offer little more than cheaper systems, and it is often hard for a new buyer to see the differences. Low-cost CAD systems include AutoCAD LT, CorelVisual CADD, DesignCAD 2D, Drafix, EasyCAD for Windows, Mini Cad, TurboCAD and Visio Technical. These low-priced packages require a less powerful computer and are four or five times cheaper than AutoCAD.

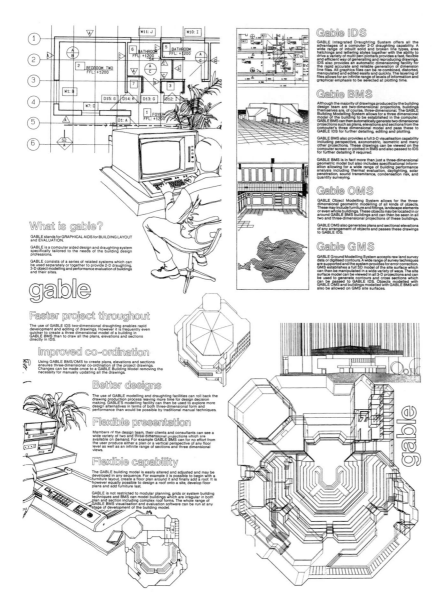

Fig. 28. System concepts of the GABLE 4D series CAD system (courtesy of Gable CAD System Ltd)

could be sorted, searched and combined with a wide range of editing and computing facilities for producing reports via a suitable word processor and with business graphics.

- The Building Modelling System (BMS) was an interactive graphical system allowing the input of building plans with user-definable material, and with surface, window and door specifications, evaluative routines and building regulation

These CAD developments inspired large architectural practices such as YRM and Grimshaws to invest heavily in Macintosh systems. This Apple Macintosh customer base then went on to include Heron, Hopkins and so on. In particular YRM[20] established a reputation for pioneering computing systems, especially the development of Intergraph (based on Digital's VAX architecture). In June 1988 YRM decided to invest £1 million in UNIX-networked Apple Macintosh work-stations on the basis that IT could be used in running and organizing the multi-disciplinary 550-strong practice in architecture and engineering more successfully.

By 1989, there were enough advances to allow the use of CAD to aid every aspect of the design process from sketch and detailed design right through to photorealistic animation. Animation and interactive multi-media applications can produce impressive presentations in colour. By 1994 Kohn Pedersen Fox (UK), Fosters, Rogers, Grimshaws and YRM were all using Intergraph software on PCs or Intergraph UNIX machines. In the UK, CAD technology has permeated the majority of drawing and design offices during the last decade. Massive reductions in the price of PCs combined with a dramatic increase in the available computational power and availability of low-cost and effective software means buying a CAD system is no longer a major issue, with the benefits of using CAD systems well documented. In 1985 the cost of equipping a single draftsman with CAD started from £30 000 and by 1997 the figure stood at £1000.

Computer graphics

When CAD systems were first developed, before the 1970s one of the key problems faced by developers was the lack of low-cost computer graphics displays. Back then computers normally interacted with users and programs via large bundles of punched cards and paper based terminals called 'teletypes'. Users fed the program and data into the computer on cards and received the results on sheets of paper. If they were lucky they might also have been able to download programs and data to magnetic tape for long-term storage. Obviously, card readers and printers were of little use to CAD developers (except perhaps those working into the field of finite element analysis). Their attentions were firmly fixed on the evolving graphic display technologies: storage tubes and vector refresh displays.

Storage tubes 'stored' an image as it was electronically painted on to a plasma screen by an electron gun — a bit like an Etch-A-Sketch.

The screen was cleared by applying a voltage across it. Vector refresh displays were a little more sophisticated. They could read vector data, store it in a buffer (normally around 64KB) and redraw it on the screen at a rate of around 14400 lines per second. Naturally, the more lines drawn, the more the screen flickered. Both types of device were generally connected to the computer via RS232 interfaces and could, literally, only draw lines. They were not cheap either. Prices started in the thousands and went skyward pretty soon afterwards. There were quotes of £12 000 for a 19-inch vector refresh monitor with 64 KB RAM in 1984.

The technology that made it possible for interactive graphics operating systems to be developed was 'raster' displays. In a raster graphics display the image is stored in the computer's memory banks as a matrix of dots, called pixels. This matrix is converted into a video signal by a Digital Analogue Converter (DAC) before display on a monitor. Because the amount of information stored for each pixel could be increased, developers were soon able to store colour information. Hence the modern computer display was born. The manipulation of the individual coloured squares or pixels from which computer and video displays are generated has led to alternative means of computer image creation to the vector-based images. Figure 12 (a–b) demonstrates this.

A typical computer screen image, for example, is composed of a pixel grid 640 squares horizontally by 400 squares vertically. When stored as pixel grids, computer images are referred to as 'bit-maps' — a mapping of the individual bits of information comprising the picture. If increased in size, a bitmapped object will always decrease in detail, as the small squares which make up its composition become larger. After enlargement images will thus appear with jagged rather than smooth edges.

Bitmap graphics are generally used when images will never be seen other than on computer or TV screens, or where complex colour or shading effects are required. Common applications include desktop video (DTV) work, the creation of high quality slides with gradient-shaded backgrounds, or the manipulation of photo-realistic images for DTP. Bitmap graphics software includes: Deluxe Paint®, CorelDraw® and Autodesk Animator® (used to create moving images). However, raster technology has its limitations. One of these is the price of the video random access memory (VRAM) used to store the pixels data. Colour raster images use lots of memory. For instance, to provide 16 colours requires 4 bits per pixel (bps) and to provide 256 colours you need 8 bps. True colour demands a massive 24 bps, providing 256 scales for each of the component colours: red, green and blue.

This means that to display 256 colours at a resolution of 1024 pixels horizontally and 768 pixels vertically (1024 x 768) requires at least 780 KB of memory; to display a true colour image at the same resolution requires at least 23 MB. Another problem is that generating the video signal for a high-resolution image at flicker-free refresh rates creates a big strain on the DAC. For instance, to generate 1024 x 768 pixel resolution image at a frequency of 70 Hz requires the DAC to process 5.5 million pixels per second. In true colour this means converting some 16.5 million colours, because each pixel is formed from three separate red, green and blue components. These figures sound awesome and indeed they are. In the 1980s, *Byte* magazine suggested such advances would be impossible.

Nevertheless, history has proved otherwise. Since 1990 the prices of VRAM and DAC have fallen dramatically — in part at least because of an increased demand for MS Windows. Graphics cards are now available for under £100 that can display true colour at a flicker-free resolution of 800 x 600, and boards which can handle true colour at 1280 x 1024 can be found for under £500. While prices may not fall much further, more sophisticated technologies are likely to appear at the current price points in future years. Furthermore, these devices are no longer 'dumb' either! They usually incorporate an on-board processor capable of increasing the performance graphics operations dramatically. The forthcoming years will see the integration of such devices and DACs at the component level, meaning that you will be able to buy both processor and DAC as a single component. Who knows how long it will be before the VRAM is added too!

CAD capabilities
Computer drafting

Symbol Libraries accumulated by the user or purchased through vendors aid the proficiency and efficiency of CAD software. These include drawings for components such as windows, doors, kitchens, bathrooms, cabinets by firms such as Pella Windows, or elements such as office products, tables, bins, lights and office furniture by firms such as Hawthorn Inc. Libraries of ready-drawn symbols including nuts, bolts, springs, screw-threads and electronic symbols take the place of stencils and adhesive transfers in traditional drafting. They can be rotated, reflected, or scaled as necessary, and copied into the drawing in the required position. One very useful graphical object is the standard frame and title box used by a practice to identify all its drawings. In the UK, engineering drawings have to comply with BS 308, engineering drawing practice for style in terms of arrowheads, leaders and the positioning and size of lettering.

Creating drawings using primitive drawing elements such as points, lines, circles and arcs or other elements such as polygons, ellipses, fillets, traces, polylines, and French curves can save time. The knowledge that technical drawings can be produced with no difficulties whatsoever allows inhibitions to be cast aside, and makes light of coping with estate roads and their circles and precisely defined radii. Drawing and editing capabilities including rotation, deletion, mirroring and multiple copying features allow changes to be made easily and quickly. Text entry and editing can be carried out in a variety of fonts and there is the ability to stretch, compress, slant, rotate, justify and edit text characters. Layering allows presentation and logical organization of data, so that the designer can distinguish different kinds of data especially when creating a complex product. Furthermore, different layers can be 'switched' on or off, increasing flexibility. Automatic dimensioning and crosshatching increases productivity and accuracy.

CAD is useful as an aid to design of buildings and forms in which there is a high degree of repetition and where repetitive space planning is common and immense quantities of information and details are involved, e.g. hospitals, schools, laboratories, etc. CAD provides an aid to reduce drawing workloads and save time in labour-intensive tasks (especially complex projects in which staff levels would be very considerable). It thereby allows feasibility studies to be undertaken and examination of options to be done quickly and easily, as well as rapid updating of drawings. Savings in time can be a direct cost saving or a revenue opportunity, depending on what you do with it. A new class of tool is replacing the drawing boards and physical prototypes to improve the conceptual design process. The integration of CAD and analytical applications allows the design to be assessed for cost, buildability or compliance with Building Regulations as it is being drawn.

Most CAD systems prove essential in two main areas.

(*a*)   As drafting tools for complex drawings. In 2D drafting it is essential to structure information well in order to use the system's ability to produce drawings at any scale, use post-processing facilities to change plotted output such as text size and font, colour, patterning and line thickness on large-scale drawings. 2D drafting exploits the fact that many buildings can best be described as a series of stacked plans. In complex geometry where projects have curved elements CAD is invaluable in setting these out easily, quickly, and with clarity and consistency of drafting style, particularly for repeat projects or repetitive elements. In automating drafting, 2D CAD carries the baggage of existing paper-based metaphors and there is therefore a need to integrate computing fully with 3D design activity.

(b) As design and presentation tools for 3D modelling and rendering, starting from the early stages of design when the model might only consist of very few elements (a single surface for the whole elevation) through to detailed design stages when complex junctions might be illustrated in 3D. As a design tool, deeper investigation of alternative layouts becomes more practical for repetitive items such as lighting or sprinkler systems.

Although most practices adopt 2D CAD drafting as a productivity aid, incorporating the computer in the design process has only happened in a small percentage of practices. Working in 2D you can not only adjust the cost of purchasing and implementing CAD downwards but also you do not necessarily lose much in overall productivity. When creating initial drawings, CAD systems are not significantly more productive than drawing boards. It is when modifying drawings that CAD systems score heavily in the productivity stakes. The difference is that when a highly accurate mathematical model of your design is created, unlike the indelible ink of a drawing, this 'soft' mathematical model can be modified very easily.

Developers of CAD systems are more interested in the time you spend thinking than the time you spend putting pen to paper. As a result, CAD systems now tend to be tailored to meet the needs of users working in different disciplines. Use of the same CAD system by large organizations, many of whom have to work together even in small construction projects, allows exchange of drawings in digital format, thus reducing the amount of re-drafting and cost. AutoCAD AEC using Autodesk's Drawing Exchange Format (DXF) is now a *de facto* standard drawing interchange in the UK AEC Industry.

However, the primary efficiency is in designing in 3D, where real benefits accrue which offer the ability to attach construction-related information to the design rather than re-drawing it in a working drawings format. In this case the ease of use of software is important, and the best applications are defined as solid modellers and the applications that allow items like doors and windows to be selected from a library are preferred.

Automatic analysis of the performance of designs

Drawing aids not only speed the process of drawing, allowing more effective use of the programs' precision, but also allow calculation of a number of relationships among objects such as areas, perimeters, moments, angles and other measurement information which can be accessed separately. Simulation tools backed by industry-leading graphics and processing power lets architects and engineers visualize models and evaluate product performance.

Computer modelling

3D presentations allow clients to envisage projects and the designer can create renditions and make changes in such projects much faster, thereby reducing the need to construct expensive miniature models. The benefits of engineering visualization go beyond client presentations (from concept, to design analysis, production and maintenance). The power of 3D graphics and mapping and imaging capabilities combined with collaborative solutions, lets users express, clarify and understand complex ideas. The technology of visualizing an artificial world in a computer virtual reality makes it possible to work on projects without having to go to the expense of erecting buildings or constructing something. The technology allows you to develop mathematical models that describe structures along with fixtures and fittings.

The 3D modelling can be analysed automatically to generate plans and elevations; extract volumetric information, and predict its response to different environmental conditions. It can also be used to design services such as ducting and cable trays accurately, as well as to produce accurate perspectives and renderings of these. In most of these cases the 3D models are more difficult to generate, so the operator needs more training. Other uses of 3D models include their use to train firefighters and other people who work in dangerous environments.

Visualization ranges from simple block modelling at the early design stage to fully rendered images for final presentation. Traditionally, the basic method of 3D-designing has involved extruding and lathing (rotating) 2D objects. These forms along with primitives (3D objects such as cones and spheres — are used to build up/create models of simple products (including buildings), while sculptural forms with 3D curves are used to create a radius within a right-angle with the aid of 3D splines.

Facilities for 3D-modelling and easily understood printouts mean that architects/engineers can design a building element-by-element and then produce drawings for clients, minimizing expensive site alterations and thereby making financial savings. Most clients have difficulties reading drawings and understanding a sense of scale, so that creating 3D renderings and animations makes it easier to communicate effectively and present designs to clients. This increases their understanding of the environmental impact and general public acceptance of proposals. There are a number of 3D programs, for example ArchiCAD®, Modelshop®, Swivel 3D®, etc., and high-end 3D programs can be used for models in drawing co-ordination while low-end programs are used for element design.

CAD/3D modelling systems cover architectural 3D computer modelling and visualization, animation, walk round/fly through, and rendering and drafting. Visualization and rendering packages include

- 3D Studio, Autovision supplied by Autodesk
- Accurender by Aztec CAD
- Alias Studio by Alias
- Alias Sketch, Infini-D, RenderZone by Gomark
- Electric Image, Strata Studio by Principal
- Extreme 3D by CU
- MasterPiece by Bentley Systems.

Several packages allow many options of computer presentation techniques and these are more easily facilitated by higher end graphics systems workstations

- Photomontaging — 2D CAD images of buildings appear superimposed on site photographs scanned from prints or transparencies, or photographed with a digital camera
- Wireline images and block massing — simple 3D wireline or cuboid block CAD models of buildings are superimposed on site photographs to show massing; the views can be obscured or

revealed from critical view points

- Perspectives can be base drawings for water colourist, or computer rendering
- 3D Settings of building set on a 2D/3D site model
- Phased development with predicted vegetation growth seen at 1, 5 and 20 years
- The Building Movie with a 3D model created allowing walk round, through, or fly over
- Desktop virtual reality (DVR) displayed on the computer screen rather than arcade-game equipment of helmets, moving platforms, data gloves and other intrusive hardware.

Higher end graphics systems workstations such as the AlphaStation 500/500 (by Digital Equipment Corp.), Model C 180-XP (by Hewlett-Packard Co.), 43P Model 140 (by IBM Corp.), Indigo2 Maximum Impact 10 000 (by Silicon Graphics Inc.) and Ultra 1 (by Sun Microsystems Computer Corp.) aim to provide more realism. A photorealistic image contains more clues about the physical properties of materials used and the quality of lighting. The key rendering algorithms for rendering realistic scenes are Radiosity and Ray Tracing.

The significance of CAD as an aid to 3D design and visualization was demonstrated at the 1993 Northern Construction Industry Computer Exhibition, when CADassist showed how a remarkable visualization of the Canary Wharf development at Docklands was 'built in a day'. While at Vauxhall Cross Development by Terry Farrell & Company[21] computer graphics have been used to explore the detailed fitout of foyers and core pods and elementary studies of building exteriors. 3D block modelling was also used to test design constraints and create an occasional realistic 3D model (in collaboration with Guiliano Zampi). The remarkable perspectives and plans of the Eisteddfod competition entry by Donald Bentley, Stephen Taylor, Andrew Houlton and Craig Muir[22] were produced using T[2] Solutions Sonata. Aluminium branches are clearly shown resting on steel trunk columns from a structural 'tree'; these branches are stressed and stabilized by diagonal steel guys. Translucent pvc-coated polyester fabric forms the envelope. There are many examples like this showing good CAD application.[23]

Virtual reality (VR)

Visualization or presentation techniques must also consider virtual reality. This, a means of measuring the environment as seen through the senses, had its historical roots in the 1920s when it was used for aviation practice. At the time, a Blue Box of simulation, with controls for the trainee pilot, was devised as a way of training pilots to land safely and without fear of causing damage to the aircraft. Although primitive, this was a start and technologists gradually added functions to it such as motion and visuals. Shadow graphs were used marking the horizon and landmarks, then utilizing slides and films. The advantages of film, while available and indeed used, were restricted to the viewing of landing operations.

The successful breakthrough came with the use of the television camera. Landscape models and computer control of the camera on a gantry proved to be effective, but a number of technical and optical problems had to be overcome within this system. For example, to keep objects in correct focus and highlight them necessitated the use of powerful spotlights that resulted in models overheating. Although this was solved by using fans, it meant that the models were more difficult and more expensive to produce. A further advance was through the use of computer-generated imagery, whereby the computer was used to create images of scenery.

Serious research into VR therefore began in the US in the late 1960s at the University of Utah, Salt Lake City, and in 1968 at MIT, Washington, where a working prototype of what has since become Autodesk's cyberspace concept was demonstrated. But it was not until the late 1980s with the key enabling technologies of low cost, high-powered computers, specialized graphic chips and the addition of sound that the first generation of VR equipment was produced incorporating the use of a CD ROM database and floppy disk.

The company which really put VR on the map is a California-based VPL Research (Virtual Programming Language) which won a NASA grant to work on VR. VPL developed the Microcosm product and its earlier Reality Built for TWO (RB2) System, which can interface with Silicon Graphics and Macintosh platforms.

In the UK, Dimension International developed Desktop VR and Desktop VR toolkit which share the same software as immersion VR but on a desktop computer, and Dimension developed SPEA GRaphics FGA 1. Developments known as 'immersive' reality immerse the subject in another world through vision, touch and sound, and there have been particular effects of successive technological innovations on refining the equipment used. An example is the development of the liquid crystal display screen, which led to the production of visette which allows the subject to see within a range of 360 degrees.

Virtuality is engaged in developing applications for commercial use in medical simulation and architecture rather than VR entertainment. V-shape is an interactive 3D modelling and VR world creation system that provides an innovative way to develop 3D content in applications such as architectural and engineering design, medical training and 3D animation. It facilitates the creation of complex objects either from within the system or imported from other modelling packages so that images of everyday objects or geometrical constructions can be assembled to create complex interactions and worlds.

Immersive VR applications can be developed on the desktop computer and offer a number of features, including a library of geometry, textures, animations and sounds. Kinematics, for example, provides sophisticated animation methods that overcome jerkiness and allow smooth-moving images to be created with skeletal animation through forward or inverse actions. 'World Creator' facilitates the creation of worlds through object selection and associated properties such as audio, motion, colour, texture and effects. VR has been defined as the ability to explore a computer-generated world by actually being in it, so that instead of looking at a screen you are enclosed in a 3D graphic universe where you can affect what happens to the virtual world as you can the real one. This definition covers a myriad of techniques and approaches, aspects of which are currently in use.

The two types of virtual reality are

(a) immersive VR — use of head-sets and in some cases full body armour alongside a proprietary parallel computer is regarded as true virtual reality

(b) non-immersive VR — just a step on from accepted visualization techniques.

Immersive VR manipulates a 3D CAD model as though it were cardboard. By the use of data gloves, a computer with parallel processing generates an image of hands which mimic the users' movements, so that a wall can be moved or pieces of the model picked up. The underlying database is automatically updated. The head-mounted display contains two high-definition LCDs which project a stereoscopic image to each eye, usually at a screen resolution of 360 x 240. There is also a head-tracking device, used to correlate head movements with the display.

The true prospect for VR is that 3D computer models and perspectives will give way to real-time explorations of the 3D design both for in-house design studies and for client presentations and planning applications.

QuickTime VR technology from Apple Computer can be used to look around a design or an existing building. As a software tool it lets you quickly build, either from photographs or computer-generated images, 360 degree virtual spaces that can be navigated through on the computer screen. You can even 'connect' rooms, allowing you to 'move' from space to space and look around spaces. QuickTime VR is different from a fly-through in that it is not 'linear' (in a fly-through you program the paths and then play-back the trip, like a video). With QTVR the computer user can choose, when viewing the image, whether to look to the left or right, up, down, stop, or move ahead. It is a navigable image.

Add-on softwares and non-CAD benefits of computer technology
Add-on softwares can help you add ductwork, plumbing, electrical service and wall details, or a complicated stairway. Facade from Eclipse can be used to draw 3D models quickly for a complex roofline. Lambert Scott & Innes, Architects, Norwich, used non-CAD softwares on a project on Castle Mill to demonstrate how the building works, and also how this would work relative to the rest of Norwich by producing photographic-quality images from design information stored in a single database, compiled from the architects' information.[24]

CAM (Computer-aided manufacture)

CAD/CAM is marriage of many engineering and manufacturing disciplines. It is computer-hardware oriented but equally dependent upon specialized software united through common databases. There are two schools of thought on how to link CAD with manufacturing. Some people believe that the best solution is an integrated system for CAD and CAM. Others believe that CAD and CAM need two separate systems.

The integrated system has all the CAD (design) functions in the same software package as the CAM (manufacturing), offering four advantages.

(a) The CADCAM system will most probably come from the vendor who is accountable for the functionality of the entire system.

(b) The transfer of data from CAD to CAM is seamless and often without any translation from one format to another.

(c) The barriers between the drawing office and the production office can come down allowing close control of the entire design-to-manufacturing process. Engineers who have the option of exploring the machinability of the part while still designing it can get the job right first time. The production engineer has the ability to go back to the CAD model and extract information needed for production, planning and machining.

(d) The production engineer has the full power of the CAD system to design fixtures, jigs and tools.

However, this integrated system has two disadvantages for production engineers.

(a) They have the ability to change something that is not their responsibility.

(b) They are saddled with the extra cost of design software whose full power they may never use.

Separate CAD and CAM systems need a good link for transferring design data from CAD to CAM. Often this is over a network, but can be a disk or tape. There are two advantages of separate systems.

(a) Both design and production engineers can have the systems that do the best job for them — horses for courses. They may have existing CAD or CAM systems and wish to make additions.

(b) Software and hardware costs can be lower for both CAD and CAM.

Disadvantages are that data transfer may not be perfect and that there is not necessarily one vendor responsible for the two systems. Figure 29 shows the place of CAD/CAM in the linear model of the design process.[25]

Over time the disciplines of CAD and CAM have developed mostly in parallel, each having different ways of defining the same thing. The result is that one set of products has been developed specifically for product design application, which is now moving towards features-based modelling with the realm of solid modelling packages such as

- Autodesk Mechanical Desktop by Autodesk
- CADkey 97 for Windows by Baystate Technologies Inc.
- Helix Modeling by MicroCadam Inc.
- MicroStation Modeler by Bentley Systems Inc.
- Solid Edge V2 by Intergraph Corp.

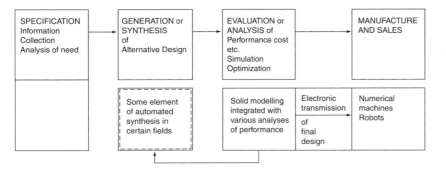

Fig. 29. The place of CAD/CAM in the linear model of the design process

- Solid Works 96 by Solid Works Inc.

- Trispectives by 3D/Eye Inc.

A second set of products, for the manufacturing group, deals with very precise features based on geometry definitions. In dealing through another system to meet various specific needs there is a lot of room for poor communication in the design-to-manufacturing process. Also, crucially in such systems, products to give meaning to the concept of integration are missing.

Although CAM is often used to define computer-aided-manufacture, the term can be all-encompassing to include production, scheduling and the like, and more often than not actually refers to computer-aided machining. These systems allow off-line programming of CNC (computer numerically controlled) machine tools. Generally these systems have geometry creation commands and, through some form of post-processing (converting the screen graphics into a language understood by CNC controller), generate a 'part-program' for CNC machine tools. This part program when run on the machine's CNC controller drives the machine to cut metal.

Six common characteristics provide a basis for understanding CAD/CAM; the interrelationship between engineering and manufacturing functions.

(a)   A computer graphics workstation (for the complete set-up for input, processing and output) is the key element in CAD/CAM systems.

(b)   The main elements of a graphics workstation, no matter how complex, include a host processor (which inputs the host application program's command data which is used to pass this data into a hardware or software routine to convert the desired object into raster format); a display controller (which allows the computer to communicate with CRT); display and output devices.

(c)   A common database (for unifying the CAD/CAM operation) grouping technology is easily entered into a central database. It

is a concept that improves productivity through the organizational common parts, problems and tasks.

(d) Software together with the means of communicating data to those who need it (whose origins lie in Adam and AD2000 developed in early 1970 by Dr Hanratty) are the basis of a host of mechanical CAD/CAM systems.

(e) Numerical control (NC).

(f) Robotics.

While the more common functions of CAD include geometric modelling analysis, testing, drafting and documentation, those for CAM include numerical control, robotics, process planning and factory management. CAE functions include automated design, simulation analysis process and tool design. Numerical control, computer numerical control and direct numerical control are all geared to performing the machining task more quickly.

In numerical control, programmed instructions stored on punched paper tapes (in the computer memory) control automatic machine tools, while in computer numerical control there is a more advanced system incorporating a detailed minicomputer or microcomputer to control the individual machine tools connected to a central computer that supplies the instructions. Numerical control is composed of four basic elements

(a)   an input interface that transforms signals into power levels that are suitable for logic devices in the control system

(b)   some form of instructions, or logic to be given to the machine

(c)   memory, which allows a machine performing a sequence of operations to know what has transpired and when to proceed to the next step

(d)   an output interface which changes the signals from the control logic into usable signals for machine action to take place.

Two of the first CAM programs (APT and PRONTO) were developed in the late 1950s, providing languages that eliminated the need for the programmer to communicate directly with a machine tool through punched-hole codes. Early CAM systems were very much language-based and ran on expensive hardware (soon made obsolete by the rapid advance of computer technology) and have largely been superseded by today's modern graphical CAM systems which are easier to use, faster, more flexible and cheaper. Windows CAM software includes

- Auto-Code Mechanical by AutoCode Mechanical, Dublin

- Master CAM by MasterCam Div of CNC software

- NC Polaris by NC Microproducts Richardson TX

- SmartCAM by Carmax Mfg Technologies

- SurfCAM 6 by Surfware Inc.

- VisiCAM Surf 5 by 3D Technologies.

Like CAD, CAM uses the Cartesian co-ordinate system (X, Y, Z). It therefore follows that anything drawn in a CAD system should be transferable to the CAM system for subsequent machining. Unfortunately, this is not true. The first problem is defining a method/medium for transfer. This involves specifying a neutral format that all CAD and CAM developers can follow to allow data exchange. Because of the rapid success of AutoCAD, Autodesk defined DXF format (Drawing Exchange Format), for saving a readable format of drawing or design on computer disk which could be read by third-party software. Although this is now the *de facto* standard among PC-based CAD systems for transfer of files, it is AutoCAD-specific and so limited to geometric entities supported by AutoCAD. Alternatively, there is the IGES standard (Initial Graphics Exchange Specification) which was created by a committee of CAD/CAM developers with the overall interests of the industry in mind.

In general, those construction industry firms wishing to exchange data between different computer systems must conform to common standards for

- magnetic media, the disks on which the exchange takes place

- communication protocols, e.g. telecom services, if data are to be transmitted

- the operating systems the computers used ,e.g. DOS, UNIX, Windows

- the applications which produce the data, e.g. CAD, word processing

- the format and structure of the data.

There are many other specifications available, but DXF and IGES are

the main two and if a CAD/CAM system does not import or export these types of files it is a serious omission. New methods of transfer have been proposed (STEP/PDES) but are not yet widely accepted. For scanned images there is the RLC (Run Length Coded) for day-to-day exchange and easy manipulation, while CCITT Group 4 Compression Standard is for long-term archiving. For printers, HPGL is the popular format and can be the fall-back mode.

The programmable industrial robot is the device that is mostly associated with CAD/CAM. It was first developed during the 1950s when George Devol began patenting his concepts. In 1958 Devol entered into a license agreement with Consolidated Control Corp., a subsidiary of Condec Corp., producing a laboratory model by 1960 and prototypes by 1962, the year Condec and Pullmann Inc. merged to form Unimation. In defining what is the ideal when looking at the automated factory, the entire manufacturing process is covered, from production to marketing to inventory to shipping.

CAE stands for computer-aided engineering and its functions include automated design, simulation analysis and process and tool design. CAE/CAM is similar to CAD/CAM except it incorporates NC tape programming. CAD/CAM/CAE-computerized design and manufacturing technology has proved indispensable for mechanical engineering.

In 1992 the UK overall CAD/CAM market was worth some £311 million, of which the architecture, engineering and construction sector accounted for £56 million. While CAD software is directed towards product design, CAM software is concerned with production control and tool design. Typical CAD/CAM users include large industries such as the auto, aerospace manufacturers of jet engines, helicopters, machine tools or products, automobiles, plant and design and layout of manufacturing facilities. Typical software for CAD/CAM includes

- Anvil-4000 (Manufacturing and Consulting Services Inc.) a new version of AD-2000

- CAD/CAM (Cadcam Inc.)

- Computervision (Computervision Corporation)

- HP-9000 Software (Hewlett-Packard Co.)

- IBM CAD Software (IBM Corporation)

- Apollo CAD Software (Apollo Computer Inc.)

- Unigraphics (McDonnell Douglas Automation Co.)

- VAX CAD Software (Digital Equipment Corporation).

CAD and CAM applications

The first commercial users in the early 1960s were car and aerospace companies such as General Motors, Boeing, Lockheed and large electronics firms, for example Fairchild and Motorola, who began using computer techniques for designing printed circuit boards and integrated circuits. Slightly later came civil engineering and architectural applications, especially for large projects in the public sector where government departments were well able to make the necessary capital investments. During the 1970s CAD spread widely into other application areas such as graphic and textile design, TV and film animation, and typography.

In the UK, CAD technology has permeated the majority of drawing and design offices during the last decade. Massive reductions in the price of PCs combined with a dramatic increase in the available computational power and availability of low-cost and effective software means that buying a CAD system is no longer a major issue. The benefits of using CAD systems are well documented.

References

1. Sutherland I. *Sketchpad; a man-machine graphical communication system*. Thesis, MIT. 1957. Also in MIT Lincoln Lb. Tech. Rep. 296, May 1965 Abridged Version in *SJCC* (1963), Spartan Books.

2. Masuda Y., Sugihara K. and Sato F. *Graphics*. Gero S. J. (ed.) *Computer applications in architecture*. Applied Science Publishers Ltd, London, 1977, 251–311.

3. Ray-Jones A. Computer development in West Sussex. *Architects' Journal*. 12 February, 1968, 42.

4. Paterson J. W. Aids to improve the designer's control of cost and performance of buildings and to speed the pre-contract and post-contract processes. W. J. Mitchell (ed.) *Environmental design: research and practice*. *Proc. EDRA3/AR8 Conf*. School of Architecture and Urban Planning, UCLA, Los Angeles, California, 1972.

5. Paterson J. W. An integrated CAD system for an architect's department. *Computer-Aided Design*, **6**, No. 1, January, 1974, 25–31.

6. Hoskins E. M. And Bott M. F. DHSS: development planning system and production documentation system. *Proc. Int. Conf. Computers in Architecture*. British Computer Society, London, 1972.

7. Meager M. A. Computer Aids in Hospital Building. *Proc. Int. Conf. Computers in Architecture*. British Computer Society, London,1972.

8. Meager M. A. The application of computer aids to hospital building. J. Vlietstra and R. F. Wielinga (eds). *Computer-Aided Design*. North-Holland Publishing Co., Amsterdam, Holland, 1973.

9. Hoskins E. M. OXSYS: an integrated computer-aided building system for Oxford Method. *Proc. Int. Conf. Computers in Architecture*. British Computer Society, London, 1972.

10. Richens P. *OXSYS: computer-aided building for Oxford Method*. Applied Research of Cambridge, 1974.

11. Richens P. *OXSYS System*. Applied Research of Cambridge, 1976.

12. Chalmer J. R. The development of CEDAR: a computer-aided design system with graphics using a central computer operating through terminals. *Proc. Int. Conf. Computers in Architecture*. British Computer Society, London, 1972.

13. Daniel P. T. A plant and buildings draughting and scheduling system, in Vlietstra J. and Wielinga R. F. (eds). *Computer-aided design*. North-Holland Publishing Co., Amsterdam, Holland, 1973.

14. Th'ng R. and Davies M. *SPACES: an integral suite of computer programs for accommodation scheduling, layout generation and appraisal of schools*. University of Strathclyde, Glasgow, ABACUS, Paper 21, 1972.

15. Roberts S. E. *The theory behind and development of an architectural sketch modelling system*. PhD thesis. University of Sheffield, Faculty of Architectural Studies, 1990.

16. Succock H. *GOAL: general online appraisal of layouts*. ABACUS Occasional Paper, University of Strathclyde, Glasgow, 1978.

17. Laing L. W. W. AIR-Q — a flexible computer simulation model for airport terminal buildings design. *DMG-DRS JNL*. **9**(3), June, 1975, 288–93.

18. Bijl A. Application of CAAD research in practice: a system for house design. *Proc. Int. Conf. Computers in Architecture*. British Computer Society, London, 1972.

19. Applied Research of Cambridge. *SSHA system*. Applied Research of Cambridge, Cambridge, 1975.

20. Practice: Computing: YRM has established a reputation for pioneering computing systems, so their decision to invest £1 million in Apple Macintosh workstations was not taken lightly. Ian Latham spoke to Computer Services Manager Ian Woodall and Chief Executive Tim Poulson. *Architecture Today*, **2**, Oct. 1989, 70.

21. Practice: Computers: Teresa Ashton on the approach adopted by Terry Farrell & Co. to computer networking. *Architecture Today*, **18**, May 1991, 84.

22. Practice: CAD: Highlights of product launches and demonstrations at last month's CADCAM exhibition. *Architecture Today*, **17**, April 1991, 92.

23. CAD: Getting the most out of the system: three case studies. *Architecture Today*, **17**, April 1991, 48.

24. Practice: Computers: Norwich architects Lambert Scott & Innes made use of new and old technology when they combined watercolour with CAD. *Architecture Today*, **12**, Oct. 1990, 92.

25. Rooney J. and Steadman P. (eds). *Principles of computer-aided design*. UCL/Open University, 1987, 3–10.

Telecommunications and networks

Introduction

This chapter is about communications systems and since the 1840s these have evolved over individual specialized analogue communications networks, e.g. sound radio (broadcast radio network), voice telephone (telephone network) and TV programmes (television network). The significance of emerging technologies, as a result of cheap computer processing and storage based on digital electronics, means that, with the right kind of transducers (microphones, loudspeakers, TV cameras and screens, modems, etc.), computers and digital communications networks can be used to store, retrieve, display, view and communicate information (in whatever format), without the need for the computer to understand their meaning. This coupling of computers with communications not only signals the convergence of telecommunications and computing but also heralds a new age.

To an extent concern is for on-line information services — Internet (including ISDN), E-mail, fax and LANs. ISDN (Integrated Services Digital Network) is an example of an emerging technology being offered by the world's telecommunications networks which combines voice and digital network services into a single medium. This makes it possible to offer customers digital data as well as voice services through a single wire at speeds of up to 64 KBPS per channel. The problem is that increased ISDN take-ups are deterred by prices and burdensome minimum contract periods largely in the hands of telephone companies.

Networks

A network is an interconnected and co-ordinated system of geographically dispersed communications devices (terminals) so that signals transmission to or among any of the devices is practical and reliable. Networks consist of a number of interlinked computers communicating with each other. They range from small LANs (local area networks) to large WANs (wide area networks). A WAN makes use of the telecommunications network to link computer systems over a wide geographical area. The computers can send and receive data among themselves and share certain devices such as hard disks and printers.

Network management concerns the administrative services involved, such as network typology and software configuration, downloading of software, monitoring network performance, maintaining network operations and diagnosing and troubleshooting problems. With a PC, modem, communications software and a telephone line, a user can also access distant information from banks, other organizations and individuals. Such a single connection is not a true PC network. This is defined as the ability of a group of PCs to communicate with each other and other electronic devices. Figure 30 shows the graphical representation of a network.

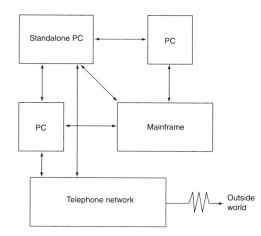

Fig. 30. A graphical representation of a network

In general, networks allow

- sharing project data between CAD workstations engaged on the same project

- sharing expensive resources and peripherals like back-up devices, plotters, CD-ROM juke-box laser printers, etc. which exploit the benefits of workgroup computing (e.g. improved productivity, distributed capabilities, co-ordinated output and collaborative working and shared information)

- passing and sharing/collecting data for central processing (i.e.

time sheets, back-up compilations, passing draft letters to and from secretarial support).

RIBA Information Technology Group says the following about networks

Server based networks have one computer acting as 'Boss' and all computers can either be daisy-chained to it or linked by direct cable in a 'star' formation. The Boss or server acts like a traffic policeman, does little else but direct data from one station to another, and upsets everybody by putting its arms down at the most inconvenient time. Because the server has to channel all the data transactions, bottle-necks occur. Usually the server has to be fast and expensive to cope! Because MS-DOS as an operating system was not designed for this type of thing, the server often runs a special operating system of its own that only system vendors, experts and trainers seem to understand.[1]

The idea of the standalone PC application is in severe decline. Organizations want word processing, spreadsheets, personal information managers and other PC applications that are pieces of the office automation jigsaw. They want to use these as bolt-in components within integrated applications.[2] An organization is no longer just automating its internal activity — it is automating its interface with the outside world via IBM/Lotus with Lotus Notes and Smart Suite; Microsoft with Exchange and its PC applications; Netscape with its portfolio of Internet products; Novell (which holds the largest share of the network market) with its excellent GroupWise 5 product; Corel +; Oracle and so on. Careful planning, selection and implementation must be considered to ensure that in the final analysis a network supports, rather than hinders, communications.

The Internet (a network of networks)

The global network of networks has its origins in 1973 as a US Defense Advanced Research Projects Agency (DARPA) venture to allow networked computers to communicate across multiple, linked, packet networks. It was created by the US Defense Department who were researching computer networks that could survive atomic bomb strikes. The Internet connected academic sites to allow them to share the few super-computers then available. However, with the network in place other uses were soon found for the system, e.g. E-mail — the ability to send text messages or files to any other user on the Internet. Messages would usually be delivered within hours — sometimes within minutes — and would then be held until the recipient read them, neatly circumventing the speed problems associated with telephone calls and the privacy problems associated with fax machines. By late 1995 the Internet was achieving critical mass, generating a positive spiral greater even than that created by the original PC.

The Internet (inter-network network) is a mesh of interconnected worldwide data networks, together carrying an entire universe of information all of which may be accessed from your own PC. As mentioned above, it developed from an academic and non-profit network linking many of the university and industry research laboratories in the defence research community in the US. Europe was then linked, and its use gradually spread widely beyond the defence community, particularly as people realized the power of electronic mail as a communications device. Research groups in many different organizations, disciplines and countries were able to communicate and co-operate, and discovered how informal information could be obtained in a way which previously only computer experts had been able to exploit.

Then the world wide web evolved, creating an incredible web of linkages between an already huge collection of information items and databases. A group in CERN (the European Laboratory for Particle Physics) in Geneva, Switzerland, and elsewhere demonstrated how a word in a document in one computer connected to the Internet (say a name in a telephone list or reference to the title of an article in a scientific journal) could be linked electronically to a previously unrelated document (say a photograph of the person or a copy of the article itself) which was stored on a totally separate computer on the Internet, often in a different country.

Many people spent long hours putting millions of these hypertext links in place for a wide range of different kinds of information in which various groups of researchers happened to be interested. WWW (world wide web) is a user-based system on a client/server architecture which is the information portion of the Internet. It is easy to access and retrieve information instantly wherever the location is held on the Internet. To begin with most web pages were static, simple, displays of text and graphics with an occasional multi-media file and a few hyperlinks. By 1997 the web was literally brimming with active, content-dynamic, interactive, user-driven displays which rely on scripts built right into web pages. Besides the www, the Internet provides many useful resources such as E-mail, newsgroups and ftp (file transfer protocol).

Newsgroups are another Internet development and describe discussion areas to which any Internet user can send messages and which any user can read. As the Internet grew to include computer systems outside the US, so the subjects covered by the newsgroups and the number of contributors to them exploded; a typical newsgroup may average 100 or more new messages per day. Newsgroups act like a bulletin board of relevance to special interest discussion groups existing on the Internet. A monitor begins a session by posting a question on a particular topic. Users then log into this newsgroup, read the question, then post an answer or additional questions. These questions and resources are accessible to anyone browsing. Most newsgroups provide Frequently Asked Questions (FAQs) and answers. FAQs usually define the group. The earliest newsgroups concentrated on technical subjects and were aimed at scientific and academic users. There are now over 20 000 publicly available newsgroups covering everything from computers to politics, celebrity gossip to missing information. The ubiquitous ALT (Alternative) newsgroups deal with some of the more exotic subjects.

A newsgroup exists for just about any subject possible from discussing the latest TV programme to comparing notes on the best PC to buy. Usenet News covers the entire gamut of subject matter and is divided into a number of categories: ALT groups cover all aspects of computing; MISC groups cover a variety of miscellaneous topics; NEWS groups cover information about the Internet itself; REC groups cover sporting and recreational activities; SCI groups cover scientific subject matter; SOC groups delve into social issues, and TALK groups engage in dialogue about an endless array of topics.

CAD users have not been left out of the information revolution, e.g. the newsgroup *comp.cad.autocad* (the name showing how Internet newsgroups are organized — alternatives to 'alt' are 'comp' for computing) contains articles from AutoCAD users throughout the world. If a new release of AutoCAD has a bug, details and criticisms normally appear first on the Internet, followed quickly by work-arounds developed by users. If you have a problem you can pose a query to one of the newsgroups; some of the replies may not be of very much use but others will contain real gems of information — Internet users can be very generous with their advice. A quick survey of current topics revealed discussions on the RAM requirements for AutoCAD Release 14, ways of getting street maps into the program and policies for billing clients for CAD hardware among many others.

The *Internet World* magazine is a useful reference publication, providing helpful hints and information about where Internet is going, how to get the most out of it and, most of all, how to get on board. Broadcasting represents yet another industry that has broadened appeal through the Internet. While conventional broadcasting merely allows the audience to hear and otherwise observe activities, a fascinating Internet medium known as 'M-bone' allows viewers and listeners to share voice, data and images over the Internet, i.e. interactive audio, video and text services with practical applications of teleconferencing and business-to-business interaction.

As a result of the nature of its development, the Internet is not highly regulated (it is self-regulating, anarchic, decentralized, federated) and there is no commercial control nor ownership. It succeeds because up to now it was designed to be (and partly has been) operated by loose federations of like-minded, well–intentioned and very intelligent people. However, this leads to important issues which have to be addressed before using the Internet as a business tool, such as security, hidden costs, the type of information, the best access methods and the likely hardware and communications costs involved, in addition to the breadth of information that you will find waiting for you. Major service providers include CompuServe, IBM, British Telecom and Microsoft; the means through which valuable information may be available.

In the 1970s the Internet was a revolutionary feat, starting as ARPnet, an experimental computer communications network with high-speed computer links and packet switching technology enabling mail and file transfer between 40 computer systems. This global network now connects thousands of computers and more than 35 million users world wide. With a minimal equipment investment, any private citizen, commercial enterprise, government agency or educational institution can chart its own worldwide course. Because it provides file transfer, remote login, electronic mail, news and other services, the Internet is considered to be the prototype of an open systems architecture. This can largely be attributed to its use of the TCP/IP protocol suite. All computers that use the Internet require TCP/IP in order to communicate and transfer data. Hence to be on the Internet, you must have IP Connectivity, i.e. be able to link other systems.

The TCP/IP protocol suite — Transmission Control Protocol (TCP) and Internet Protocol (IP) — was developed in 1974 as part of the DARPA Project. TCP/IP was an important technological development because it paved the way for standardization and inter-operability. TCP/IP has achieved status as the network protocol of choice — the *lingua franca* of the Internet. Leading computer hardware vendors including Digital Equipment Corporation, IBM, Hewlett-Packard and Sun Microsystems have incorporated TCP/IP support into their product strategies. Microsoft Corporation has also included TCP/IP support into the latest version of Microsoft Windows and Windows NT. One reason that TCP/IP has been so successful is that it is an extremely practical and reliable solution for heterogeneous, mixed-platform open systems environments. The use of the TCP/IP protocol enables many different computer systems to connect, whether they are Windows-based PCs or UNIX-based PCs, Macintosh computers, Windows NT servers and UNIX-based host computers. All use the TCP/IP protocol as the *de facto* communications standard for Internet connectivity. TCP/IP is also the preferred networking standard for organizations migrating from proprietary mainframe environments to an open systems model. In addition to the popularity of TCP/IP, several other factors have contributed to the Internet's rapid-fire growth. They include the proliferation of client/server applications, powerful networking technology, the evolution of standard open-systems and the widespread use of the UNIX operating system.

Electronic mail exchange has generally been the most popular Internet application, followed by the ability to read and post services (for queries or comments) in newsgroups and mailing lists.

Notwithstanding a number of major unresolved technology and system issues such as security, privacy, bandwidth and network reliability; for architects and engineers there are two main benefits of being connected to the Internet

(a) access to current data is made available over the net by suppliers, design agencies, financial service agencies, consultants, interest groups and so on

(b) opportunity to show off their wares and thereby make information about an organization available to other net-users. This can be achieved by designing a home page — an electronic page of data which will act as a shop window for the organization (company profile and details, a portfolio, awards, approach and so on) consisting of text, photographs, sound, video

drawings and covering several pages or screens of information. The data will reside on the computer of your IAP (Internet Access Provider). Specialist companies (Archinet and Pringle & Colyer) offer a full web service with Archinet linking some leading UK practices, including Foster & Partners, Ferry Farrell, Chris Wilkinson, Mott MacDonald and Morris Associates.

Engineers have always been great users of computers and have access to very powerful machines which make getting involved with the Internet relatively easy, especially if an intranet already exists. One of the main attractions of an intranet (see Glossary) is that the technology is cheap. Most organizations introducing an intranet will already have staff using networked desktop PCs sharing files across a high-powered server computer. Some will already be using groupware such as Lotus Notes and will be used to communicating electronically.

The server computers which form the cornerstones of an intranet are based on open technology and common standards, allowing connections to all the different types of computer on the network. A PC user can view and access a document no matter what kind of computer or software created it or where it is actually stored. In addition, an intranet can handle sound and video files, live video-conferencing, complex graphics, even animation, all accessed through the browser. The technology which drives an intranet is also extremely intuitive and easy to use without much training. Each piece of information on the intranet is joined to another by hypertext. Employees use the links to skip from item-to-item, just by clicking on where they want to go. To make it easier to find the right information, the company can attach related documents or files to a particular home page. The intranet can transparently deliver the immense information resources of an organization to each individual's desktop with minimal cost, time and effort. Companies already use intranets for a range of tasks such as storing and disseminating masses of data which would normally have to be printed for distribution to staff, circulating software upgrades, providing marketing and sales support, circulating training materials and timetables and providing employees with access to internal knowledge databases.

Alternatively, powerful search engines can trawl the entire intranet and even go out into the wider Internet and bring back the right information. Sensitive corporate information can be protected from the outside world by security software known as firewalls which requests a password and asks for other forms of identification.

With access to thousands of networked computers worldwide, possibilities for the mutual exchange of ideas and collaborative co-operation are limitless. Other uses include transferring files, logging on to a remote computer, or simply exploring the world. Finding information on the Internet is based on the principle of client/server architecture. The server is the computer system that contains information, such as electronic mail, database information or text files. The client system requests the appropriate data from the server system. While a client system can travel anywhere in the world it must ultimately attach to the server, whether to transfer files or to log on to another computer system.

The Internet provides a fast, inexpensive, way to communicate interactively with one or multiple users across the globe. Figure 31 shows the benefits of connecting to on-line information services in order to access databases, news, E-mail, etc.

Traditionally, Internet users have relied on character- or text-based UNIX public domain software commands as navigation or research tools to access remote computer systems, but learning the right UNIX commands is not always an easy feat. Since the Internet was designed for UNIX technical users, even the names of computer systems can be intimidating. One such UNIX command, designed in 1991, is 'gopher', the University of Minnesota's Newsgroup hound, public domain software that provides a menu-based format for roaming the Internet. Gopher is called that because it 'goes for' information and because the college football team is called the Golden Gophers. It resembles the www in that it provides a way of linking together different information sources.

While the www provides links in the form of 'hot buttons' on pages of styled text with pictures, gopher takes the form of menus and submenus on each server, some of which lead to other gopher servers. To use gopher, however you must know and type in the address of the gopher system you are trying to access. Another Internet software program that enables users to access remote computers is 'telnet'. Provided you have an account on the remote system or know the name of the public username/password pair, any client can telnet to any remote server. To download software patches for debugging or testing, ftp (file transfer protocol) enables you to transfer files across the network. Files can consist of software, text or graphics, and files to download are usually located in directories on ftp servers. Publicly available search and indexing tools — WAIS, Veronica, Archie and Jughead — can be used around the world to locate the particular server that has the information you require.

Architects/engineers can connect the computer installations to the telephone network to benefit from on-line data, electronic mail and other services. They can benefit from on-line information services (on the Internet, WANs, or LANs, etc.) precisely because it is difficult to keep the office/technical library current and containing all information. PC users who can purchase on-line services can key into information resources that are often not available elsewhere. Therefore time is saved because the search time is a fraction of that necessary to complete a traditional search. Using a regularly up-dated information database (some of which may be specialized and specific) held somewhere else on a large computer via modem/PC is the alternative. Architects/engineers can therefore scan databases for British Standards, Codes of Practice, British Regulations and so on. In-depth information on CD-ROM subject by subject in colour allows relatively easy access to Building Regulations (fire safety, resistance to the passage of sound, conservation of fuel power, structural stability, access for disabled, etc.) and all other reference material needed when undertaking a project. Examples include Codes of Practice, legislation, governmental advice (including BRE information), guidance and recommendations from specialist organizations as well as the increasing

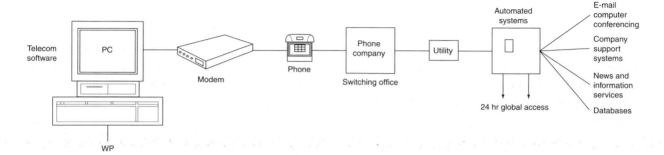

Fig. 31. Connecting to on-line information services

number of building products and manufacturers. Searching by company, series, title, product, etc. will produce a list under each heading which can be printed out. Many suppliers and firms are already offering CD-ROMs of their products in preference to catalogues.

These services allow pooling of resources across offices (in different regions, countries or continents). Designers can have access to super-computers to simulate and analyse designs before expensive time-consuming physical models are actually built. Super-computers and their high-performance desktop workstations and servers increase the number of options that can be examined and analyses that can be pursued, not only maximizing flexibility but also allowing computing budgets to be used in the most cost-effective manner. Furthermore projects can be shared between offices — allowing the spread of the so called 'sun never sets' design offices — according to workload, expertise and so on. Drawing work can be farmed out to countries which can offer cheaper labour rates, without incurring any delay. On-line services enable individual architects to work with other architects and engineers at remote locations and to take on larger projects than they would as sole practitioners (Could this be the demise of the individual architects or engineers? We need to wait and see.)

Electronic communications can reduce the number of people needed on site, saving time and money. Site management can be improved by using E-mail, video links, wireless networks and electronic ordering, and the use of bar-codes can help keep tabs on material stocks to streamline the supply. Video conferencing and shared screen working or teleworking allow complex problems to be discussed. Different disciplines can collaborate interactively in the computer environment.

The key to success in the information age will be management. As the number of registered users of the Internet passes 26.2 million, sending the sheer quantity of electronic data available soaring into the stratosphere, the issue becomes not so much how information is stored, but how to find it.[3]

In general, on-line services can provide the fast accurate service required by clients. Using the Macintosh and QuarkXpress, Adjini of Aldo Rossi's firm edited *Aldo Rossi: Architecture 1981–1991* and *Aldo Rossi: Drawings and Paintings for Architectural Press*. The firm uses (AOL) America On Line's electronic mail facility to transfer files between associate architectural firms. By storing names and telephone numbers, taking notes and communicating via fax and E-mail (which is done by hand-held computer when on the road), Newton Message Pad is used to manage client information, save sketches, send faxes and generally to ease the burden of managing internal projects and keeping in touch with the home office and clients across multiple time zones.[4]

For the US situation, Grabowski[5] discusses various BBS (Bulletin Board Service) services — both commercial, general-interest and private — and the Internet. Bulletin Boards are electronic equivalents of noticeboards. Commercial general-interest BBS services include CompuServe, America On Line, Prodigy, Delphi, GEnie, and eWorld while private limited-focus services include Architecture On Line, AE&C Info Net, AIA On Line.

- AIA On Line offers CBD (Commerce Business Daily), directories for architects and consultants, employment referrals (job listings), MasterSpec tutorials, NIBS guide specifications and ASTM abstracts.

- Architecture On Line is Online Architectural Journal, a product of Princeton Architectural Press and an architectural information resource.

- AE&C InfoNet offers access to the CBD, A/E industry information, ACEC bulletins, Harper & Shuman bulletins.

- AE Resource offers access to CBD, CAD and graphics files, product demos.

- ReproCAD Network is a long distance reprographics and file exchange information with access to service bureau in the US and Canada.

- SourceNet offers access to Canadian government corporate, real-estate jobs and product information.

In the UK there is the Quantarc electronic library from Poulter Communications. Running on a standard workstation, it provides access to 25 000 pages of product literature stored on CD from more than 9000 manufacturers. There is also an immediate fax facility. The technical information service on CD-ROM from RIBA services contains comprehensive information from the Technical Information Microfile, providing enhanced use for practices using computers in the preparation of project documentation.

Another service, Building On Line, consists of 18 databases on Viewdata with information on BCIS (Building Cost Information Service) prices, planning applications and tender invitations, Building Regulations and determinations, British Standards, BRE publications, contract forms, agreement certificates, contract case law and programs for structural and thermal calculations. Building On line database shows how Construction Law Online can give access to the latest contract law decisions using data prepared by the research department of James R. Knowles.

Contract law is a rapidly changing field in which the importance of up-to-date information is obvious. Use is made of contract law decisions and contractual issues covered in 40 or more law reports and legal and construction journals, with the cases indexed by subject and contract clause.

On Demand Information, a development of BCQ systems, gives access to product literature, Building Regulations, British Standards and full-text technical information. Whereas BCQ is based on CD-ROM, On-Demand Information is accessed from a central database via an ISDN telephone link. Information on the database is updated daily and a single line allows access by up to eight PCs and the system is said to be quick in operation, with an eight-page colour brochure appearing on screen in about 90 seconds. The advantages of greater speed and storage capacity of CD-ROM over traditional fiche and microfilm systems have meant that full Construction & Civil Engineering Index (i.e. Technical Indexes) is now available on CD-ROM.

Electronic mail: E-mail

As late as 1980 the term 'electronic mail' was still being used as a generic term for a number of mail room and messaging systems, including fax, telex and mailing and labelling machines. It is now reserved for messages that are on a screen from an intra-office network or, more commonly, from outside the office and over the Internet. This is also often shortened to the term E-mail.

By attaching documents and graphics to an E-mail message, users can eliminate phone tags and faxes. Joining mailing lists allows users to receive specific information on a selected topic whenever something new is introduced. Vendors of E-mail services use a computer as a centralized letterbox and subscribers to the service compose a message on their screen. When ready they telephone the computer and send the message through their modem, quoting the code (address) of the recipient. The recipient also has to be a subscriber (and have a modem and computer). The message is held in the computer memory until the recipient, using a password, telephones the computer and downloads the messages (known as 'looking in the mailbox'). If desired one can use the E-mail vendor to forward a message to a telex subscriber. One E-mail vendor offers to send messages anywhere in the country. This is achieved by a subscriber sending a message to the vendor who sends it by fax to a courier, who delivers it by motorbike.

Until a few years ago E-mail was simple. Messages consisted of text and, since they were almost entirely within the organization, it was easy to control the software that created and sent them, and the way staff used them. Processing E-mail was simply a matter of reading it, then passing it on or throwing it away.

E-mail has become multi-media, with the exchange of digital images, attached files such as spreadsheets and presentations, MIME-encoded messages (mail messages comprising sound/graphics sent over the net) and HTML web pages. All these bulky, different and complex data types can be a handling and processing problem and in some cases force organizations to upgrade the amount of disk space used to store messages. However, network systems such as GuideWare (*www.guideware.com*) contain intelligent agents that track E-mail, logging who reads it, who deletes it, who replies to it, and can if necessary automatically E-mail the sender with these details. Also with the variety of data types now sent, dedicated E-mail packages designed with text in mind may no longer be adequate, so that in-boxes need to be integrated with a tracking package and a database for storage and retrieval.

E-mail has now become so ubiquitous that any number of applications can produce it, and not simply dedicated E-mail packages such as Eudora or CC-Mail but networked groupware, bulletin board software like AT&T mail, and even proprietary Internet browsers such as AOL. Increasingly people are sending and receiving E-mail from portable phones, not just computers. According to figures collected by the US-based magazine, *Electronic Mail & Messaging Systems,* there are more than 150 million E-mail users worldwide. Until 1996, the number of Europeans adopting E-mail was doubling every year, and by 1997 it was still growing at a rate of closer to 50%.

Almost every BBS (Bulletin Board Service) and on-line service can offer you some kind of gateway into the Internet for electronic mail (E-mail) — it is the lowest common denominator service. The standard way of addressing Internet mail is *person@organization.something.something*, but the way you send mail out from a local system varies from system to system. To send a message from the Internet to a CompuServe account (70007,1254), you would send to '*70007.1254@compuserve.com*'.

One of the most commonly asked (and understandable) questions about the Internet is 'how do I find a person on the Internet?'. There are at least ten different ways of doing this, none of them particularly reliable. It is hardly surprising that finding people is so tricky, since there is no central authority with which every user registers to get on-line. The population of the Internet is growing incredibly quickly, and companies and individuals do not want their electronic mail addresses listed.

E-mail offers advantages, including that of speed, enabling peripatetic employees such as sales staff to receive messages each evening in, for example, their hotel bedrooms. Messages can be sent to a large number of people very quickly and easily, which can either improve productivity or reduce it if the system is abused. It is a way of gaining access at all levels — even busy executives have been known to respond to E-mails more frequently than returning phone calls. Disadvantages include not knowing when a recipient will actually read the message, and security problems.

Facsimile, telefax or facsimile telegraphy: fax

Fax refers to the method of sending and receiving graphics and text over a telecommunications network, by converting visual images to electronic signals and then converting them back again into a copy of the original. The medium was invented in 1842 by Alexander Bain, who originally used a pendulum in his scanning device. The pendulum was replaced with a photo-electric system, and commercial services using this system were available before 1910. Fax machines are referred to by groups, ranging from the earliest ones (1, 2, 3...) to the later ones, according to improvements, e.g. speed of machines (group 1 takes six minutes to transmit an A4 sheet of material, or four minutes with degradation of copy, group 2 takes three minutes, group 3 takes up to a minute). The machines can also 'talk down' (group 2 can transmit to group 1, group 3 to 1 and 2).

Desktop fax services offer faxes sent directly from the user's PC

rather than from a dedicated central machine, demonstrating clear cost and efficiency savings. However, documents can be sent inadvertently to the wrong recipient, for example, which could breach confidentiality, or details may be altered in error. Fax machines print on thermal paper or through xerographic or ink-jet printers. Fax disadvantages include an inability to handle large volumes compared with other transmission systems. There is too the annoying problem of junk mail by fax, which both ties up the machine and costs the recipient for thermal paper, and also allows transmission of bogus messages.

Local area networks: LANs

Beginning in the early 1980s a new type of network emerged, referred to as LAN (local area network), a high-speed database that connects a variety of computer-based systems together on a 'party-line'. It can only operate locally up to a maximum of ten miles. A LAN connects several small office computers in one organization using cable in order to move data (text, graphics images or voice) between desktop computers, workstations, a mainframe computer, other input/output devices and even a data PABX, thereby sharing a facility (database printers, etc.).

Successful workgroup printing requires a high degree of commonality at the LAN interface and the recognition that most customers live in a multi-vendor environment. A clearly defined printer strategy is important and requires investment and management. Decent printer management can bring costs down, generating a full range of benefits. It is more efficient and simple to tell the printer to produce the required number of original prints, all of the same quality, rather than printing the document then walking to the photocopier to produce reductions with all the attendant delays, loss of quality and likely problems.

A LAN therefore allows

- the sharing of hardware such as printers, plotters and disk storage
- the sharing of centrally stored information/data and use of all types of software
- communication between individual users.

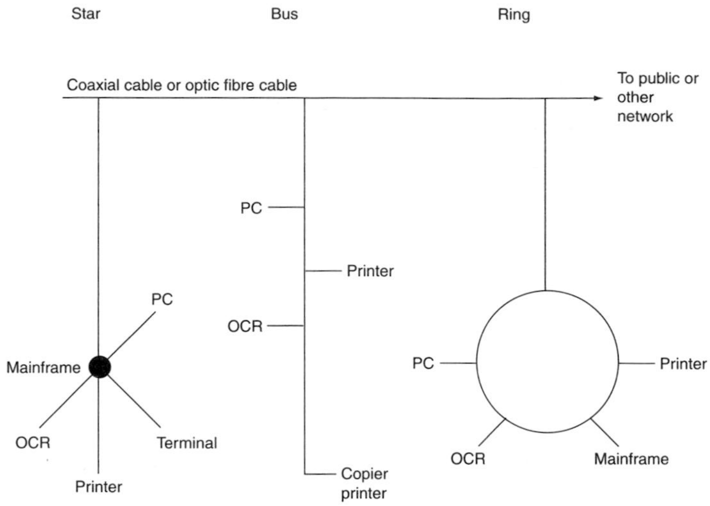

Typically a LAN consists of

(*a*)   a server (software on hard disk or hardcard)

(*b*)   the network interface (usually on another card)

(*c*)   transmission channel media (e.g. pieces of wire (i.e. twisted-pair wire), coaxial cable or optical fibre)

(*d*)   a switching centre (also known as a 'controller') which co-ordinates and controls the network and the data flow.

Furthermore, LANs are distinguished by

• Shape (star, a ring or a bus) i.e. LAN typology as in Fig. 32. In a bus network system all stations or computer devices communicate by using a common distribution channel, or bus.[6] While in a star network all stations radiate from a common controller; in a ring network all stations are linked to form a continuous loop or circle. Figure 33 shows how each of the three basic LAN architectures may be interconnected to form a WAN (wide area network).[7]

• Categories (general-purpose or special-purpose, baseband or broadband). The base difference between the broadband and baseband is capacity. Baseband carries a single transmission at any one moment without disruption. Broadband splits the bandwidth into channels, allowing more than one transaction to occupy the line at a time. Baseband works best in LANs while broadband is more reliable and better for outdoor instal-lations. With broadband, data facilities are capable of handling frequencies greater than those required for high grade voice communications, i.e. greater than 300 characters per second.

• Sharing or arbitration (token-passing ring, token-passing bus and CSMA — Carrier-sensing multiple access).

A major disadvantage is security. LAN evaluation should prioritize considerations for speed, data, volume and types of devices supported. Networking software products include Network (Novell); MS Net (Microsoft); PC Net (IBM); Tapestry II (Torus Systems). CLAN (Cordless Local Area Network) allows computers to communicate with

Fig. 32. LAN typology

each other and with peripherals (printers, etc.) without the need for complex and expensive cabling and other infrastructure, such as the raised floor to provide for cable trays and cable managed systems furniture required to allow distribution of IT in 'intelligent' buildings.[8]

The emerging technology of the new cordless telephones such as Ericsson's Freeset Cordless PBX, based on the new European DECT Standard for mobile phones (designed specifically for use within buildings) allows the location of up to 1000 cordless telephone extensions in a building. The handset provides all the functions of a traditional telephone and is not tied to a specific desktop.

The benefits of a cordless PBX include the ability to place business calls direct to the required person wherever they are in the building, thereby increasing efficiency and reducing outgoing call costs. The cordless local area network for portable computers signals another leap forward. IBM, AT&T, Olivetti and Xircom have all launched PCMCIA cards. These credit card sized units slot into notebooks and portable computers and can connect from wherever the user is based within the building into a central server, as well as to printers and faxes. They can also move, being passed seamlessly from cell-to-cell as the user moves around the building. More importantly, the portables can create small peer-to-peer networks where, say four users can automatically share the same document, work on it simulta-neously and then depart with an up-to-date version each. This creation of small, fluid, work teams reduces the need for extensive paperwork and allows all users to have the same information or

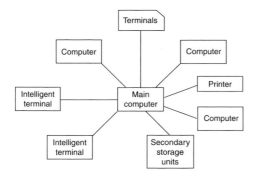

(a)

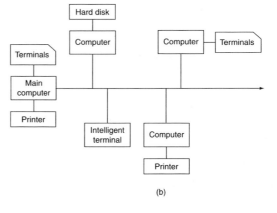

(b)

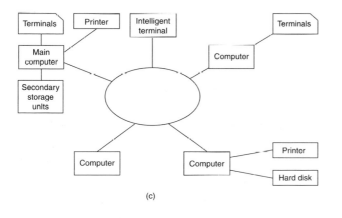

(c)

(d)

Fig. 33. The three main architectures interconnected to form a WAN (wide area network): (a) star; (b) bus; (c) ring; (d) WAN

access central resources such as databases during meetings. True 'hot desking' can be realized where both people and communications technology are free from a fixed address.

References

1. RIBA Information Technology Group. *PC Networks*, 8, November, 1992.

2. Meeks B. N. The wired society. *Byte,* **13,** 8, August, 1988, 138.

3. Thompson C. J. Information age: The next generation — if the pundits are right we are about to enter a new era in computing. *Building*, June 2, 1995, 45.

4. Beaubois T. Macintosh and architecture: a special supplement to Architectural Record. *Architectural Record*, 5/1995, A4.

5. Grabowski R. The Profession: Computers in practice: Finding a needle in the on-line haystack. *Architectural Record*, March, 1995, 40–41.

6. CAD and IT. *Building Design,* No. 1229, July 28, 1995, 41.

7. Redmill F. *The computer primer*. Addison-Wesley Publishing Company, Wokingham, UK, 1987, 25.

8. Tangney B. and O'Mahony D. *Local area networks and their applications*. Prentice Hall Intnl (UK) Ltd, Hemel Hempstead, 1988, 3–5.

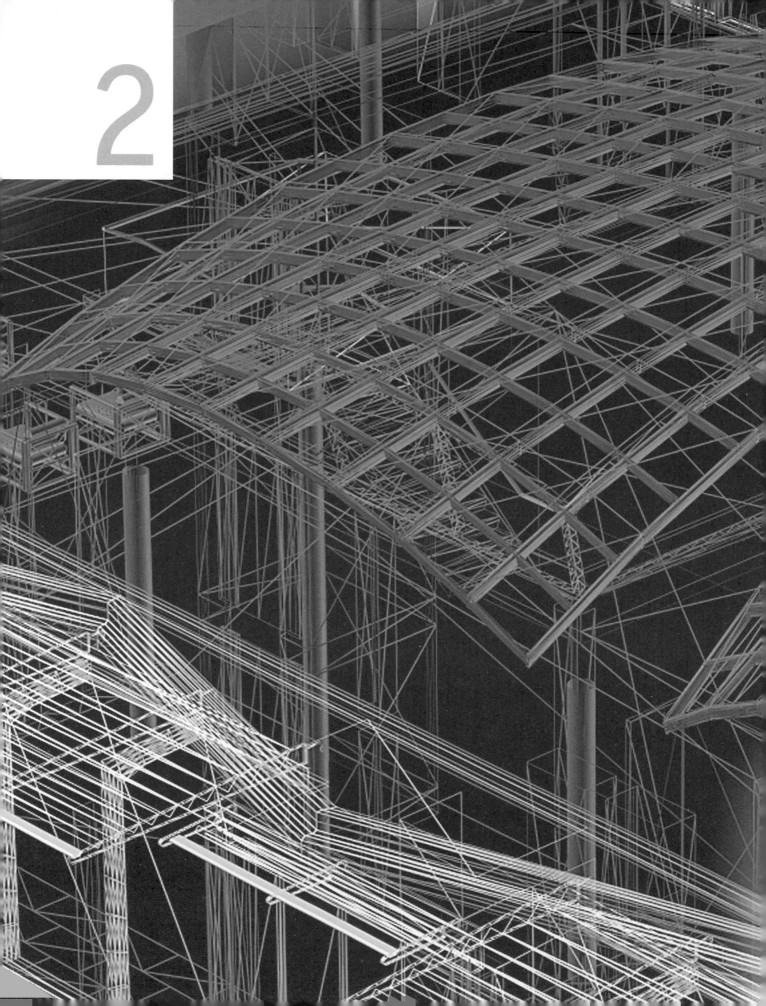

2

Part two | Case study organizations

Introduction

This part examines a number of case study practices looking at the differences in resourcing and organization for IT. Of importance are issues of the size and structure of the office, the way the practice works, the relationships and communication between principal and other staff and the role of clients in the process, all of which are inexorably intertwined.

Choice of case study organizations

The firms chosen for detailed study cover

- different sizes — e.g. small, medium, large, very large, etc.

- different structures and ownership patterns — e.g. sole proprietor, private partnership, limited company, etc.

- different services offered to clients in very different markets — e.g. architectural and/or engineering design, design-based consultancy, multi-disciplinary, etc. The services reflect diversity as well as specialization in particular areas such as commercial and industrial buildings, educational, medical and residential construction facilities

- different philosophies, practices and working methodologies.

No attempt is made to suggest that the firms represent a statistically relevant sample, or a comprehensive range of possible types. There are omissions, such as those from the public sector, e.g. housing, educational, postal and transportation facilities. Nor are they randomly chosen. They were selected on the basis of knowledge of the firms, the respect that they command within the architectural and engineering professions and the likelihood that their histories

would shed some light on how IT changes have taken place and have affected their processes. The chosen firms appear to show enthusiasm for the use of computers and new technology in their daily operation — for instance using computer techniques equivalent to drawing and physical model making and as design tools.

The diversity of case study organizations reflects

- single-principal practices dependent on the reputation of one person and this individual seems to take a hand in every project and consequently rarely runs one

- federated practices sharing resources and some staff

- larger practices that are partnerships in which the partners all lead for particular jobs or take corporate roles

- practices configured more along corporate lines with each partner playing a particular and defined role, taking responsibilities for certain stages of work rather than individual projects.

Consideration of the case study organizations is not done on the basis of prioritization but in alphabetical order to reflect different design practices and strategies. Also each practice has different IT policies, resources and operational characteristics.

Study methodology

The study methods covered

- interviews with principals (and staff where relevant) about the firm as a whole

- practice profile and marketing information made available by the organization or from the Internet location

- selected projects identified with the co-operation of the case study organization.

Interviews were undertaken based on a checklist of minimum desirable details relating to each firm. The object was to gain an insight into the working of the practice rather than a focus on the design achievement. Questions sought to discover trends and main issues. The information involved looking at the following.

(a) *Background,* i.e. a brief history of the practice covering the organization structure and how it relates to the office adminis-tration, to projects and CAD management. The main question is how is the practice organized and resourced to exploit IT? What are its policies on IT? Are there any IT standards? The background is also important in showing the development of IT tools within the firm. For example

Many of the computer-aided architectural design systems that have been developed were produced for use by specialized organizations, and are quite specifically oriented towards a particular building type. Typical examples are Harness hospital design system and SSHA housing design system.[1–3]

(b) IT applications for *text manipulation, databases, specification writing,* e.g. Microsoft Office for text manipulation, letter-writing, report-writing, standard forms, databases and specifica-tions; QuarkXpress for desktop publishing work. In addition this means examining other software used, e.g. Sage for accounting; the 'Project Command' for time sheets; 'No Teamate' for spreadsheet or Specification Manager for specification writing.

(c) *IT applications for CAD,* e.g. AutoCad and MicroStation for all aspects of work, e.g. all working drawings, 3D modelling, animations, presentation plotting; for specialized rendering

software Intergraph's Modelview. The main emphasis is to show that CAD is more than AutoCAD but relates to the description of the practice's complex process as an idea is developed from concepts to sketches to 2D to 3D production information and/or rendering and video for client presentation.

(d) *IT applications for Telecommunications and Networks,* i.e. hardware — types and number of machines (e.g. Apple PowerMacs, Pentium PCs, etc.) including type of network and links with other offices or consultants as members of the project/design team; and a description of software including network software type and other considerations and future plans. Also of importance is the nature of integration with the client base and other professionals.

(e) *IT training within the practice,* i.e. the nature of training, help desk and trouble-shooting provisions, e.g. whether this is through in-house using experienced users, vendor support or consultancy. Discovering what are the main IT issues of concern to the practice? For example rate of change of technology, incompatible data formats, training, legacy, etc.

(f) *Selected projects* seeking to show innovativeness and highlighting how useful IT has been. They are important reference material on IT applications by the Practice. The projects are described by details: location/site plans, floor plans, elevation/section, details and computer images. These are not meant to be comprehensive but to provide examples of how the practices use IT tools in day-to-day operation and from project to project. Also when looking at projects computers should not take all the credit for the design. (Note, the figure numbers refer to the corresponding text in the selected projects sections for each case study.)

References

1. Mitchell W. J. *Computer-aided architectural design.* Van Nostrund Reinhold Company, New York, 1977, 93–99, 101–104.

2. Bijl A. *Ourselves computers — difference in minds and machines.* MacMillan, London, 111–112.

3. Bijl A. *Computer discipline and practice — shaping our future.* Edinburgh University Press, 1989.

Alsop & Störmer, Architects — 1

Background

Alsop & Störmer was inaugurated in 1979 and has two principals (William Alsop and Jan Störmer) and five directors. The practice has electronically linked three offices (London, Hamburg and Moscow) with a total of 73 staff, 51 of whom are architects. The London office is multi-disciplinary employing architects, graphic designers, landscape architects, quantity surveyors, urban and interior designers.

On the practice's design philosophy William Alsop notes

The office has a completely open mind about what architecture is. This is not a closed question as far as we are concerned. There is a continuing exploration of form, colour, functional, social and behavioural issues. These investigations resolve themselves in buildings and structures that offer a richer experience to both the user and visitor. The client is considered as an integral part of the design team. This is necessary in order to establish a base which will allow the project to step beyond the expectations of all involved in the design process.

The strength of our architectural product derives from research for innovative results, simplicity for flexibility in use and economy in cost and technology for new solutions while respecting tried and tested methods.

Alsop & Störmer has managed its own in-house Quality Assurance Scheme since 1992 with each project progressed in accordance with set guidelines subject to client scrutiny.

The practice's work method entails several management procedures.

(*a*) Project team structure defining responsibilities, authority and interrelationships.

(*b*) Design procedures in which design packages are developed to ensure meeting of

   i)   client requirements in terms of brief, budget, engineering construction

   ii)   relevant regulations

   iii) safety, maintenance and durability requirements

   iv) appropriate interface with suitable manufacturers

   v)   development of component design prototypes

   vi) proper recording authorization and confirmation of design changes

   vii) suitable recording of informal communications.

(*c*) Checking procedures to ensure compliance with technical and other requirements.

(*d*) The project architect is the team leader for any project. The regional government centre Hotel du Departement, Marseilles was the 'breakthrough' building for the practice. The current range of work covers transport interchanges, corporate, civic, arts and heritage, leisure, retail and residential building types.

The practice's process involves definitive concepts at the early stages of the design providing the broad overview and generated via 2D paintings by William Alsop and developed using a combination of 3D physical and CAD models.

IT policies have evolved over time and are now based on making everyone use computers. Peter Angrave is the director responsible for IT within the practice with Christopher Egret CE the CAD manager co-ordinating with the consultancy of Ian Martin Associates for advice and day-to-day support. Although planned with IT needs are based on project requirements and the practice's policies, there is no separate IT budget.

Alsop & Störmer's computer hardware and electronic communications have been expanding — 38 Apple Macintosh CAD workstations (ranging from Quadra to Power Macs) with MicroStation and 3D software; 8 Apple Macintosh administration workstations; 2 PC (486) workstations with AutoCAD 12 software; Hewlett Packard Electrostatic Colour A0 plotter and A1 plotters; E-mail addresses (London and Moscow) and web site, 2 ISDN lines. Computer software is typical — consisting of MicroStation, AutoCAD and

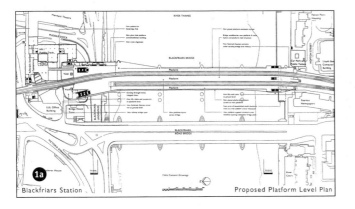

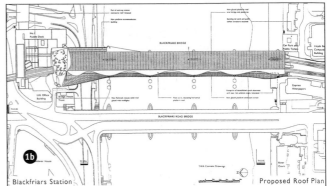

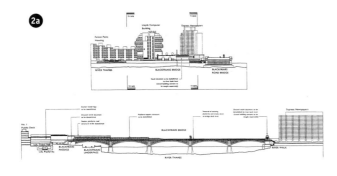

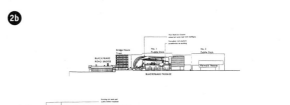

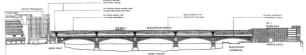

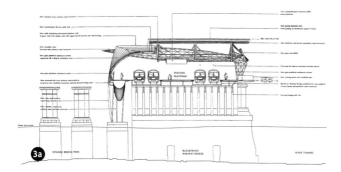

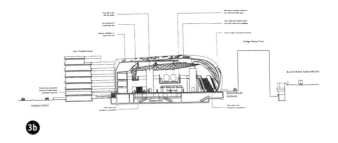

MiniCAD; Presenter Professional; Digital Darkroom, Model Shop II, Pagemaker, QuarkXpress, Photoshop; Microsoft Project; Microsoft Word and Excel.

IT applications for text manipulation, spreadsheets, databases and specification writing

Alsop & Störmer uses Word for word processing (architects type their own letters assisting the two dedicated administration support staff typing project work, marketing information, contact lists, etc.). Excel is used for drawings issue, graphics, tables templates. QuarkXpress/Photoshop are used for DTP reports and other presentations.

Other software includes Microsoft Project whose use is to be expanded. Specification writing is done via NBS on disk depending on the type of project and client requirements. Arena software is used for accounts by the two dedicated people who make up the accounts section. The section is also responsible for entering timesheet information on to their database once this has been done manually.

IT applications for CAD

The firm uses MicroStation (beta copies and the full version) and Strata Studio. Also available are two copies of AutoCAD on PCs (486) to meet specific client requirements. Within the London office CAD is mainly for 2D layout drawings and 3D modelling of concepts for massing structure and other exploration. The need and desire of the practice is to relate 3D models to drawings in MicroStation.

Typically the process in the practice starts by using MicroStation proceeding to the use of Photoshop/QuarkXpress. A fair amount of sketch details are still done by hand. This process represents an interaction between computers and freehand in accordance with the design progress. The office CAD manual explains some of the office standards and routine procedures.

IT applications for telecommunications and networks

Alsop & Störmer have one administration server and two CAD servers on an ethernet network. ISDN lines allow direct links with printing organizations, branch offices and clients. One dedicated E-mail workstation allows receiving and sending of information otherwise use is made of floppy disks. Drawings of current projects are stored both as hard copies and archived weekly on DAT tapes into specific project archives to facilitate easy retrieval.

The website location *www.alsopandstormer.com* is important to the firm as a means of communication because much of the practice's work is overseas. The website is being designed in-house and will consist of sketches, paintings and photographs.

IT training

The firm's IT training policy involves a mixture of formal and informal methods being aimed at openness and flexibility — what is best for the project and responding to demands being two of the important issues for the practice. Also experienced users help out with basics and there is an emphasis on learning by doing. Two to three training sessions for 2–3 people are often arranged including a number of presentation seminars. The practice has never had a complete overhaul of all the systems and therefore not needed *en-masse* training.

Other than concern about flexibility one other important issue for the practice is that of interfacing using various formats.

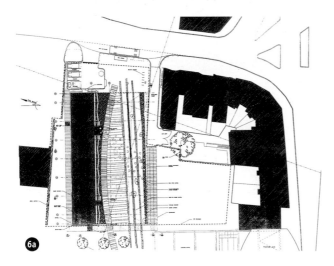

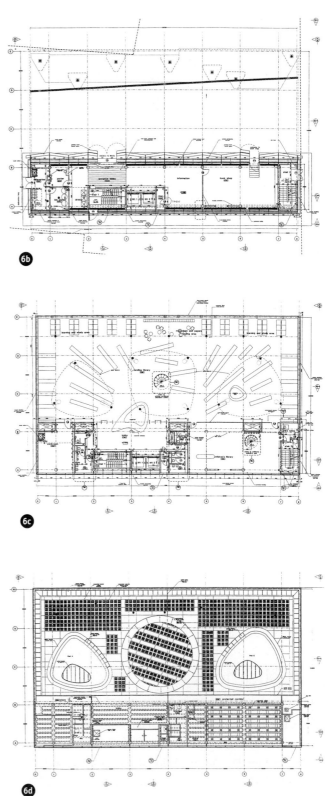

Selected projects[1]

Blackfriars Station Thameslink 2000, London, 1997

This project uses computers because of the shapes, form and size
of project including a dedicated CAD Quality Plan incorporating in-
house CAD standards submitted to the client (Railtrack plc).

1. (a) Proposed platform level plan (b) proposed roof plan
2. (a) Existing sections and elevations (b) proposed sections and
   elevations
3. (a) Typical cross section (b) proposed North Bank
   section/elevation
4. Photomontage of view from South Bank river walk towards
   Blackfriars passage

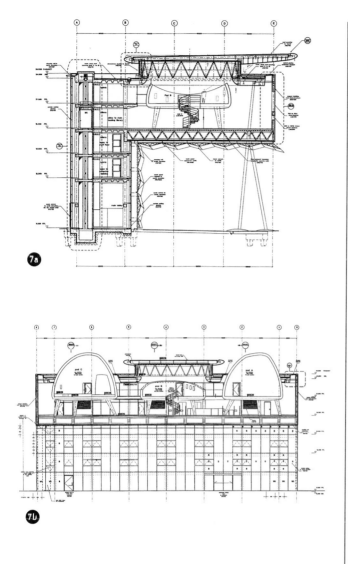

7a

7b

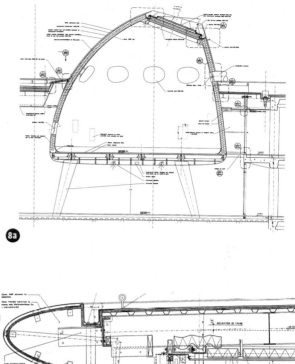

8a

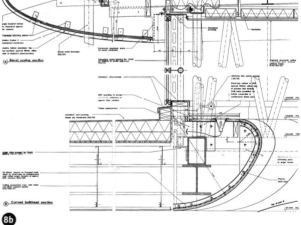

8b

Peckham Library and Media Centre London, 1997

In this project setting out was facilitated by the use of computer.

5. Painting of the building — an early conceptual form by William Alsop
6. (a) Site plan showing the hard landscaped pedestrian access (b) ground level plan (c) level 4 plan (d) level 5 reflected ceiling plan
7. (a) Section A–A (b) section B–B showing the pods (A, B and C)
8. (a) Pod A section (b) beret key details produced using MicroStation

La Fregate Restaurant, Jersey, 1997, in which the £500 000 cafe resembles an upturned hull[2]

9. Initial painting by William Alsop
10. (a) Ground floor plan (b) east and west elevations (c) service volume setting out of surface skin made easy by using a computer
11. South elevation

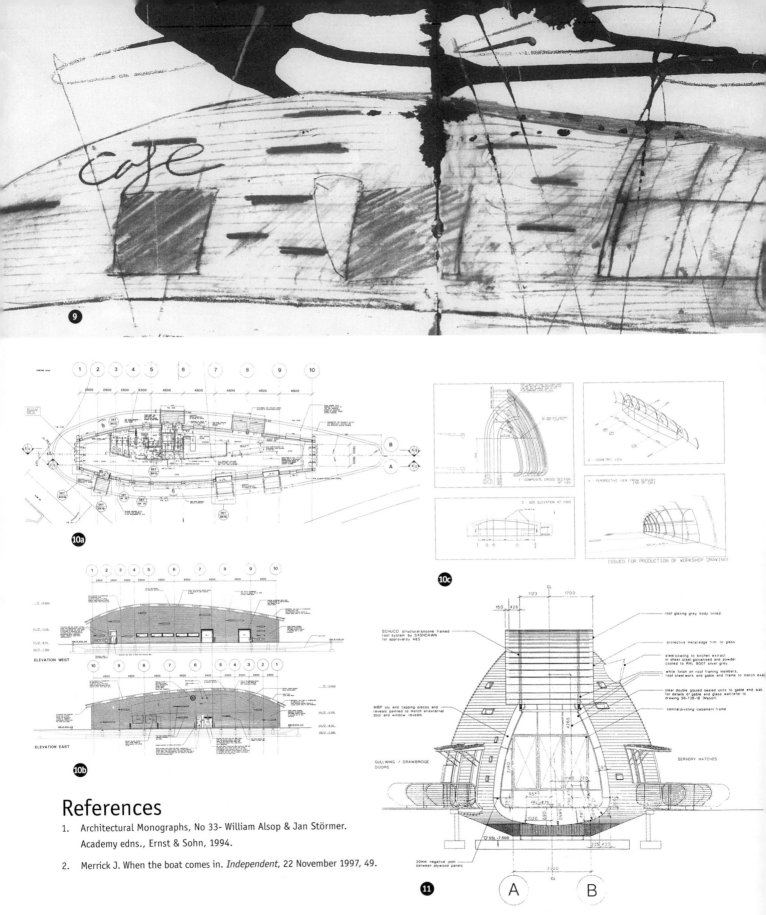

References

1. Architectural Monographs, No 33- William Alsop & Jan Störmer. Academy edns., Ernst & Sohn, 1994.

2. Merrick J. When the boat comes in. *Independent,* 22 November 1997, 49.

Anthony Hunt Associates, Structural and Civil Engineers | 2

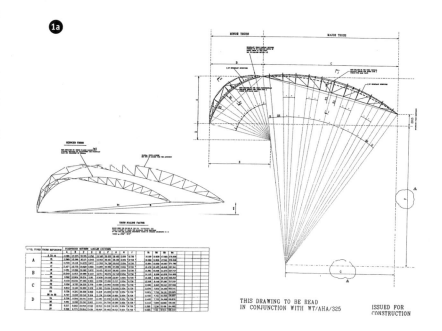

1a

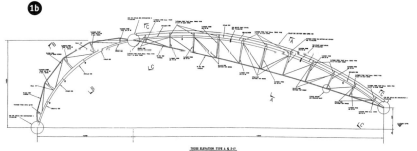

1b

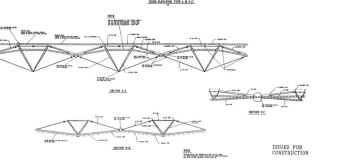

Background

Anthony Hunt Associates was set up in London in 1962 by Anthony Hunt as a structural engineering practice with contacts such as Team 4 Architects. In 1976 the practice become a partnership expanding into civil engineering in 1986 and moving out of London to Coln Sl Aldwyn and then Cirencester.

By 1988 the firm joined the multi-disciplinary building design group YRM Partnership Ltd, merging with their civil engineering business to form a new operating division with over 140 professional, technical and support staff and known as YRM Anthony Hunt Associates. In April 1997 with the collapse of YRM Group, Anthony Hunt Associates Ltd re-established itself with 6 directors including Anthony Hunt and about 50 staff in offices (London, Sheffield and Cirencester). The directors who set the practice policies are responsible for associates and senior engineers in charge of projects.

Anthony Hunt Associates has worked closely with many notable architects (in UK, France, Middle East, USA) in the development of the design, with fellow consultants and with contractors to ensure integration of architecture, structure and services. The range of buildings includes landmark projects such as Reliance Controls, Swindon 1965; Willis Faber & Dumas building, Ipswich 1975; Sainsbury Centre for the Visual Arts, 1978; the Waterloo International Terminal, 1993 and the Law Faculty, Cambridge 1995.

The practice has experience of working with conventional materials but has become specialist in less usual technologies such as tensile membranes and sophisticated glass engineering. The firm is well known for 'attention to detail' both in the exploitation of alternative solutions to problems and for the importance it places on the elegance and efficiency of design details (for example the use of rods and nodes of the typical high-tech structure). The practice is keen not to be seen as specialists only in steel design.

Typically, the practice sees itself as being involved as early as possible in the design and construction process , 'backroom boys' support to architects and project team from the iterative geometrically-driven scheme work stage, through detail design and production information stages.

Mike Purvis is the IT manager for the practice based in the Cirencester office. He is responsible for overseeing IT systems and standards (e.g. fonts, set-ups, levels, presentation controls, etc.) and at the moment for responding to day-to-day queries. He points out that practice formats and ways of doing things need not come from the top but can be driven bottom-up and from any of the offices.

Anthony Hunt Associates bought its first computer, a Hewlett Packard 45 for £20 000 which it used for structural analysis on the Inmos (UK) Ltd project, 1979. In 1988, as one of the five divisions within the multi-disciplinary group YRM Partnership Ltd, the practice had access to sophisticated shared resources, including computer-aided design, drafting and technical support services. At the time YRM was the testing bed for Intergraph Europe, suppliers of MicroStation.

IT applications for text manipulation, databases and specification writing

Anthony Hunt Associates uses Microsoft Office; Word for general administration, letters, project submission documents, reference lists,

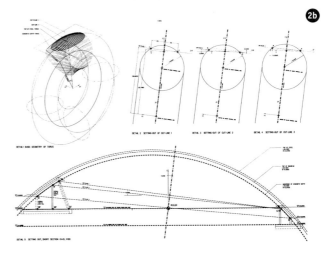

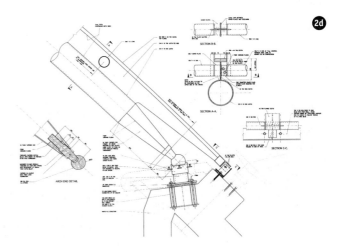

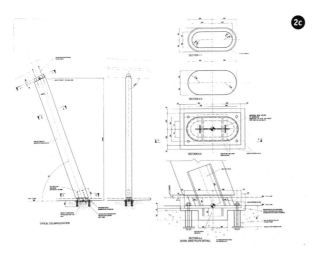

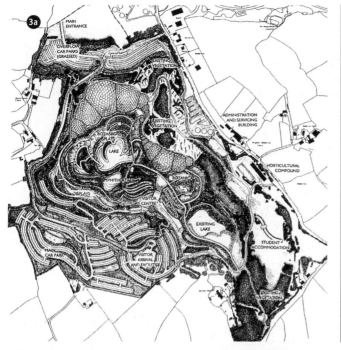

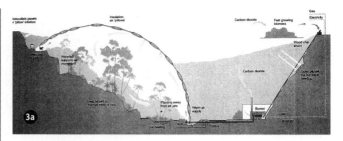

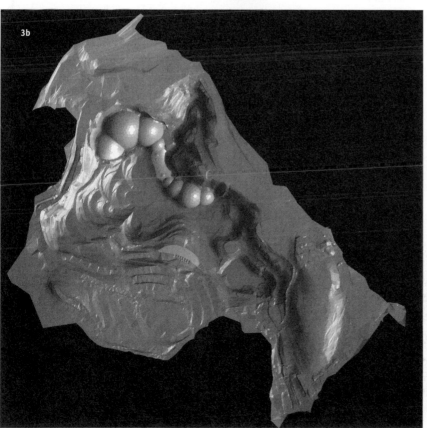

memos and desktop publishing purposes; Excel for calculations, issue sheets and pro formas; PowerPoint is new but is to be used for presentations; Access is to be used for future company databases. Word is also used for specifications.

Aldus Pagemaker is used for submission documents. File Maker Pro is used for the current office database including job historical information, costs. Timesheets are done manually before electronic input by the accounts section who use Sage Accounts software.

Other software includes that for structural analysis (Staad, Strap, Multi-frame and Scale) and a number of proprietary analysis software. The practice does not tend to write software although subroutines have been produced.

IT applications for CAD

Anthony Hunt Associates uses MicroStation and PowerDraft on a new NT4 network. The practice has 12 CAD workstations, six analysis work stations (Pentium 200s), six administration workstations (Pentium 166s) and one laptop available as a management resource tool.

All the hardware and software was replaced in 1997, the result of the reorganization and the establishment of Anthony Hunt Associates as separate from YRM Group. All the Apple Macintoshes and the UNIX Network were discarded and data transferred on to the new network except the marketing database which has still to be transferred and re-defined. During reorganization considerations were made to turn to AutoCAD. The merger with YRM also meant that the practice has only one copy of AutoCAD.

IT tools are used extensively in most projects such as the Eden Project — a series of linked geodesic

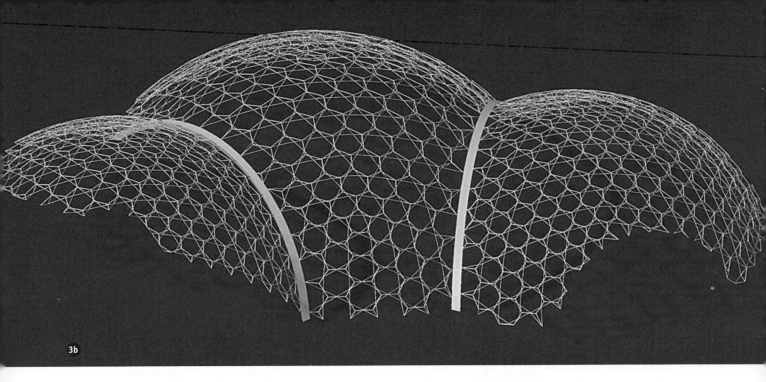

3b

domes — organic in shape along the cliff face sited on the uneven ground. To the practice

the project is one of the most exciting the firm has worked on and because the ground is fractured, the geodesic solution is the lightest possible and one which allows off-site fabrication.

Other software used by the practice is the In Roads Terrain modelling and Intergraph software — an add-on to Bentley's MicroStation but it is expensive. Finite element analysis capability is being developed and is being used more and more.

IT applications for telecommunications and networks

Formerly Anthony Hunt Associates had a UNIX network which has been replaced with NT4 with two servers (one in each office) and ISDN links to all offices. E-mail links are available using BT Internet. Security is ensured by full back-up every day and four weekly archive for tapes.

IT training

The practice has a policy of continuous in-house training with external courses provided by a local company for anything new and desirable. The use of in-house experienced users is also very valuable. There is a reliance on vendor support.

The main issues of concern to Anthony Hunt Associates are

- the need to make the best use of IT including disseminating IT knowledge within the practice without creating a dedicated and sophisticated IT department or set up

- the need to keep up-to-date with both software and hardware requirements including improving accessibility for staff from 50% to desired level of one man per machine

- recognition that CAD modelling should be seen as aiding the design process with any specialist modelling being seen as separate and therefore to be outsourced or paid for by the client. IT does not necessarily save time but cuts out thinking time, owing to the tendency to draw everything on CAD without appropriate evaluation. However, IT can increase the level of operating confidence such as the ability to deal with current design codes and solid geometries

- the need to be aware of what Alan Jones, director refers to as 'black boxing' i.e. losing the sense of a 'good practical feel of the structure' or the intuitive checks of high value in a structural engineering practice. One way in which the practice has been addressing this is by way of having crit. sessions as part of quality management.

Selected projects

Waterloo International Terminal in London, the National Botanical Gardens in Wales, and the Eden Botanical Gardens for The Eden Trust (with Nicholas Grimshaw & Partners) all show the importance of IT for Anthony Hunt Associates when involved in major projects.

Waterloo International Terminal Project

IT was immensely useful in the roof design, a process driven (both in terms of CAD and structural analysis) by dealing with the complex geometry. The roof of the new train shed involved complex geometry which also altered along its length. The introduction of a 'universal joint' enabled standard glass panels to be used otherwise there would have been a high degree of customization required. Discussions with the contractor introduced the idea of using castings rather than fabricated metal components.

1. (a) and (b) Analysis of truss types, truss support details and development of 3D details were enhanced by the computer applications

National Botanical Gardens of Wales

This project is situated on the site of Middleton Hall, Swansea, a large landscaped garden with many water features, completed in 1815 by Joy Paxton and abandoned since 1931. The £20 million scheme includes the restoration of three lakes and walled gardens together with the introduction of several new buildings aimed at educating visitors. Other facilities will be housed in the refurbished frame buildings.

The centre-piece of the scheme is the great glasshouse which is sunk into the crest of the hill and capped by a 100 x 50 m clear span glazed dome roof in the form of a toroid. IT applications enable geometrical descriptions of the structure, analysis of information from CAD to Staad, setting out and on-site prefabrication indicated by the precast column and steelwork details. The main building houses plant displays based on the various climatic zones created from regions of the African continent. An in-situ concrete bowl forms the base to the dome whilst the structure of the glass roof is a series of slender steel arches restrained by the integral cladding supports. The entire roof is clad in a double glazed system of flat panels, trapezoidal in shape, held in a nominal and discreet aluminium framework.

2. (a) Great glasshouse: the centre-piece of the project

(b) great glasshouse geometry

(c) great glasshouse precast concrete details

(d) great glasshouse steelwork details

Eden Botanical Gardens

The project for The Eden Trust is situated in a china clay pit (up to 70 m deep cut into the escarpment over many years) near St Austell, Cornwall.[1,2] In an effort to respond to fabrication, transportation and erection techniques, the design is based on advanced proven technologies and on the geometry of spheres with a system of straight tubular compressive members connected by standard cost connections. The £100 million scheme about sustainable agronomic development covers an area of around 8 ha with 4 ha of internal space within a 15 ha disused china clay pit. The lightweight structure of minimal surface area and maximum volume is made from steel tube, and 'glazed' with ETFE (ethyltetrafluoroethylene) foil which will be inflated with air warmed by photovoltaic panels to provide insulation. The foil which is only 0.5 mm thick and offers 97% transparency, is tough enough to be walked on and is self-cleaning. A visitors centre will introduce arrivals to the project before they set off on a journey through three biomes (environment spaces) where they will experience climates taken from the Mediterranean, desert and tropical rain forests.

The biomes wind around the side of the quarry providing clear spans of up to 120 m and heights of 60 m (some 10 m taller than Nelson's Column). The use of digital terrain modelling was of importance for remodelling of the pit, based upon early surveys by mining engineers, to provide access and car parking areas. The complex steel frame provides the structural support whilst the envelope is formed of the 9 m diameter ETFE foil cushions (less than 1% of the weight of glass) inflated to give them rigidity. The floor of quarry is below natural ground water levels so a comprehensive water management scheme is required, including recycling of water through the waterfall in the rain forest.

3. (a) Site plans and cross-section (courtesy of Nicholas Grimshaw & Partners)

(b) model images of the biomes in the former china clay pit

References

1. Pearson A. Green Hows. *Building Services Journal*, April, 1998, 44–5.

2. McDermott C. *Twentieth-century design*. Carlton Books Ltd, 1998, 397.

Aukett Associates, Architects, Interior and Landscape Designers | 3

Background

Aukett Associates is a multi-disciplinary practice formed in 1977 by founding partners, including Michael Aukett (who was positively disposed towards Information Technology and its application within the practice). The firm is a large practice of about 140 staff with one main office in London and a branch office in Glasgow. Aukett Associates Europe is a group of partnerships in several countries (Belgium, France, Germany, Ireland, Italy, The Netherlands) combining professional skills and clients. The group seeks to offer the ability to act locally together with a commitment to provide a consistent quality of professional service throughout Europe.

The practice has expertise in a number of areas — architecture, masterplanning, urban design, landscape architecture, structural engineering, space planning, services engineering, interior design (605 fit-outs such as Heathrow Compass Centre) and graphic design. These services can be provided independently or in combination to suit specific client preferences.

The range of work covered by the practice includes corporate headquarters, business parks, technical centres, research laboratories, hotels, retail and leisure, transport and residential areas. No speculative work is carried out and the practice is committed to integrated design, bringing in all primary design disciplines, so that multi-disciplinary project teams can be created in close proximity using common CAD models, project procedures and management structures. Projects are led by co-ordinators responsible to directors.

Harry MacCourt is assistant director responsible for IT provision

①

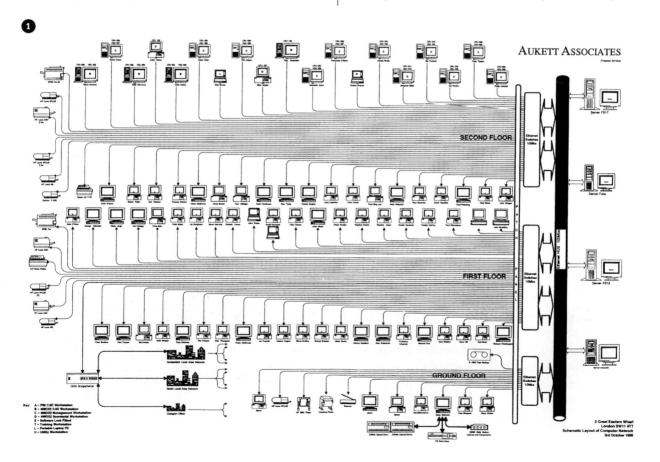

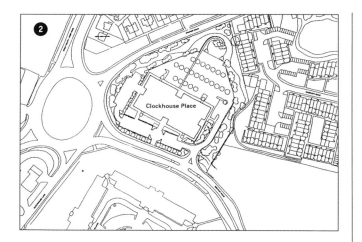

within the practice. Richard Worthington, Aukett Associates' computer manager since 1991, formerly worked with the Greater London Council as an architect before going into IT in 1981. Messrs. MacCourt and Worthington see their role as providing the tools for architects or engineers or project groups to do their work better. The

long-term strategy is to make these tools readily available to improve the practice capability (from 50% computer fluency, 30% computer literacy and 20% novices). The main idea is to have every computer on every desk running all the software thereby allowing seamless integration.

A brief history of computing within the practice shows that the practice had its first computer in 1978 — a terminals-based system PDPII (by Dec) which was used for accounts. It was selected because of its number-crunching aspects and often for printing out of reports. There was no word processing capability. By 1981 this computer was replaced by the Burroughs B25 (supplied by Converging Technology Software). It represented more of an early office machine and an early networking system which subsequently grew to approximately 48 workstations linked by a mega-stream cable — the high proprietary system was still in use until 1992.

1983 saw the introduction of CAD RICUPS a minicomputer supplied by GMW Architects. This system was available for five years, only to be replaced in 1987 (after making hard copies of all the data for

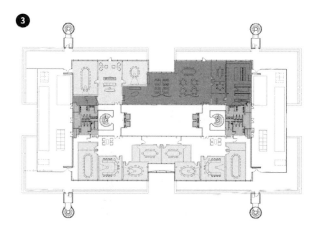

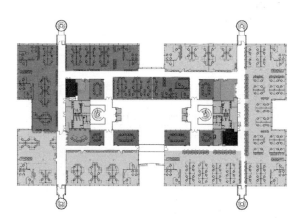

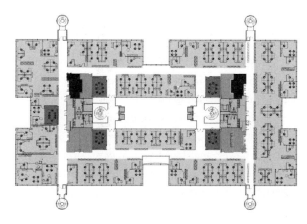

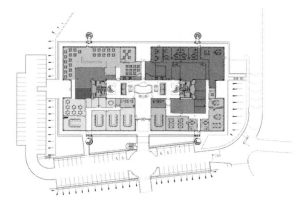

archiving) by PCs running AutoCAD. In 1990 the practice bought two Apple Macs for its graphics department only as a system most suitable for the specialist area. During 1991–2 an ethernet server system based on the old-thin wire was adopted for the network and the initial two servers grew to four. This network was upgraded in 1995 to the star type structured CAT UTP. This corresponded with a move from the Embankment offices where the practice had been located for five years to the present premises with a completely new infrastructure. The schematic layout as at October 1996, shows the nature of the computer network (see Fig. 1).

In order to keep up with the technology and competition and to enhance business efficiency the practice replaced the entire kit in August, 1997 — a feat made simple because the practice leases all its hardware. This has the advantage that replacement every three years enables a rather flat cash flow (e.g. £1000 per person per year). This might appear wasteful in that 486s, P90s and Pro200s were replaced even though being capable of running quite well.

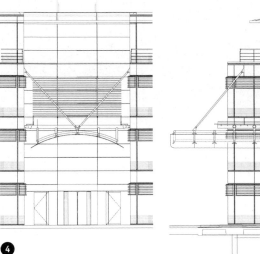
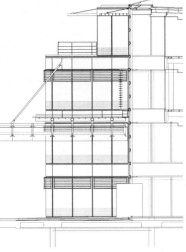

4

IT applications for text manipulation, databases and specification writing

Microsoft software is used for administration purposes. Word is used for word-processing activities; Excel for calculations for heating, cooling, electrical loadings and so on, and for drawings and document issues. Two copies of Access are used for a photographic scanning library for archives.

The practice has a customized specification generator called 'Enigma' based or developed from the NBS. No use is made of the NE Specification. NBS system and project management software with a library of standard components or details using the RIBA-CAD digitized technical library have also been used.

The practice has designed sophisticated in-house software both to monitor project profitability (project time/cost analysis, expenses) and staff levels and commitment and for marketing purposes. The in-house document management program, Timesheet, is linked to a central database, Phoenix, based on Foxpro. It acts as a File Manager system and allows one to view and modify resource requirements (cost centre profiles, consumables, number of prints, etc.) and to generate invoices.

Other software includes Windex, Hevacomp, PowerPoint, for structural engineers Integer, Sand, Oaysis, Fasteel, and for accounts Sun Account on NT4.

IT applications for CAD

Aukett Associates got into IT in the 1970s when it bought a computer to help with accounts but CAD implementation allowed extension of computer use in the practice. The 140–160 strong practice now works extensively on screen for producing mainly production information and limited 3D modelling. The graphics department uses Apple Macs to run QuarkXpress (with better picture-handling), Photoshop and Illustrator to produce reports.

Architects and engineers work on PCs running AutoCAD 14 and one copy of 3D Studio Max and Visio to carry out visualizations at the client's request and for limited 3D construction detailing. The in-house drawing register (file manager type system) program is a front end add-on to AutoCAD 14 rewritten (from DOS 16 bits to DOS 32 bits) with reference to AutoDesk's Mechanical Desktop.

Quality Manual has been available for three years, e.g. layering to BS standards. Waivers and aspects, such as all drawings issued are signed by hand, are means devised to discourage 'laziness' a consequence of using computers.

⑤

IT applications for telecommunications and networks

In the practice computers, telephones and faxes can be plugged in anywhere in the office using CAT5 UTP network which is significantly faster than the traditional ethernet, the network used by many practices. The CAT5 UTP network is marshalled using Novell and an ISDN link between offices to transfer drawings. The firm has direct ISDN links with some of their major clients, for example Asda, Dixons and Glaxco. The practice also has high-volume plotting, colour A0 plotting and scanners.

A strategic problem for the practice is the legacy one. The practice would like to develop an intranet as soon as possible.

IT training

Aukett Associates included training for all its staff in the contract when it chose Bristol-based supplier AZTEC CAD to supply its software upgrade, which included moving to AutoCAD 14 from 12 and replacing Windows 3.1 with Windows NT. Richard Worthington worked with AZTEC to devise a series of training seminars. These included 12 AutoCAD seminars, graded to suit users of differing experience. Experienced users got three hours of training, while less experienced users had four hours and novices received three days.

Training was done in groups of up to six staff. Those people with more AutoCAD experience went on the early courses. Instruction on the new system was done in groups of up to 20. 'We wanted to get a basic understanding for every single member of staff' says Richard Worthington. Although included in the tender price, AZTEC's courses averaged £400 per person per day. The total training content of the tender was around £10 000.

Richard Worthington notes that the practice has carried out limited specific training in 17 years apart from when it is upgrading systems.

I usually train newcomers myself. If they are complete novices on CAD and want to learn they have to make some effort themselves. The practice will pay for courses but they have to do them in their own time either as night classes or full-time in holiday time.

This ensures that staff who are keen to learn IT will quickly reveal themselves. He prefers staff to address IT queries to him rather than asking the person next to them. The practice's IT training budget makes up 6% of its IT budget, which for 1997/98 was £400 000

6

because of the cost of the new software.

Richard Worthington sees the issue of bandwidth as the main problem and major bottleneck with the Internet especially because many architects use the Internet for transmitting CAD files. In this case it is cheaper and more convenient to send a file on disk without any file size limitations or Data Exchange Format problems such as requirements for the necessary ISDN kit at £1500 a time. The practice has, however, developed ways to deal with such problems, for example version control and the fact that files are 'off-line' if they are being used on site.

Another issue is the need for more integration of software to allow a more seamless operation. There is also the issue of practicalities and culture barriers to full integration including simple things such as filling timesheets manually and passing these to accounts although the capability is there for electronic transfer.

Selected projects

These projects indicate how the firm works extensively on screen for producing both 2D and 3D production information and 3D modelling.

Clockhouse Place Bedfont Lakes

This project for the Hanover Property Unit Trust was a £16 million 'cutting edge' example in the speculative office building. With 10 146 m² of office accommodation for 850 staff the building achieves almost 90% net/gross floor area efficiency.

1. Schematic layout of computer network at Aukett Associates' London office
2. Site location plan shows Bedfont Road entrance
3. Ground, first, second and third floor plans. The entrance elevations and cross section show the 7.5 m x 7.5 m and 9.0 m x 7.5 m structural grid and 1.5 m planning module
4. Canopy
5. 3D images of the building
6. 3D images of the building

Autoglass Headquarters Building, Bedford (3600 m²)

This building was designed to reflect Autoglass' business (clearly glass to be a prominent material) and create a building which broke away from the traditional corridor and office feel to provide an environment that was open, light and created a sense of the business being together.

7. Ground and first floor plans and cross section show the office layout designed around an open plan arrangement, a feature required by the client to foster teamwork relationship (the 7.5 m x 7.5 m and 9.0 m x 7.5 m structural grid).

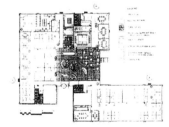

GROUND FLOOR PLAN

FIRST FLOOR PLAN

CROSS SECTION

7

Battle McCarthy, Consulting Engineers | 4

Background

Battle McCarthy is a multi-disciplinary consultancy providing building services engineering, structural engineering, urban design and specialist environmental design within the built environment, concentrating on the production of high quality, integrated, low energy buildings.

The partnership was formed in 1993 by Guy Battle and Chris McCarthy both of whom previously worked with the research division of Ove Arup & Partners. On the one hand Guy Battle is an environmental and building services engineer with experience in a wide range of low energy projects ranging from housing to office buildings. On the other hand Chris McCarthy is a structural and civil engineer with over 18 years' experience as a design and project engineer. Battle McCarthy Partnership is actively involved in education lecturing at the Bartlett School of Architecture, the Architectural Association (AA), Royal College of Arts (RCA) and the University of Bath, and writing regularly for *Architectural Design* magazine, *World Architecture* and other publications.

The practice has two offices located on Poland Street in London, one an administration office and the other the main workplace incorporating BMES workshop. BMES (Battle McCarthy Simulation) is a research wing of the practice carrying out research into building physics in order to develop a better understanding of the impact of environmental, climatic and physical forces on architecture and urban forms. BMES seeks to use computer and complex interactive calculations to help design buildings for many different moments of time enabling these to respond to changes in the ambient environment.

The type of projects carried out by the partnership range in scale from single houses to headquarters office buildings and complete new urban settlements. The engineering philosophy of the practice is based on (*a*) the primary goal of engineering 'to maximize the use of materials, energy and skills for the benefit of all' and (*b*) on several engineering design objectives — for instance to maximize human comfort, efficient planning, to design for change, minimize running costs and energy consumption, maximize usable space, minimize capital and maintenance costs, and protect and enhance ecological values.

The strategy of the practice is to aim to bring a holistic engineering approach into the design process and to develop a greater understanding/communication within the project team. The practice has worked with leading architectural practices such as Alsop Störmer, RHWL, Terry Farrell & Co., Richard Rogers

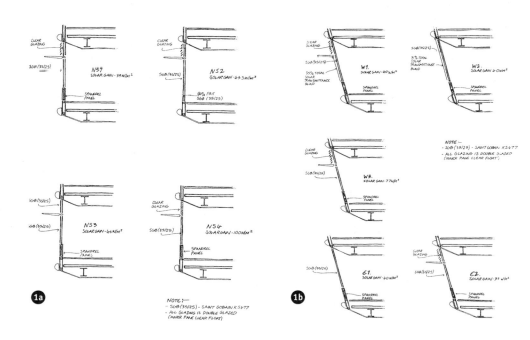

LIGHT SHELF AT 2.2 m

Daylight factors as a % using Pilkington St. Gobain Cool Lite K Green Glazing / Clear Float double glazing

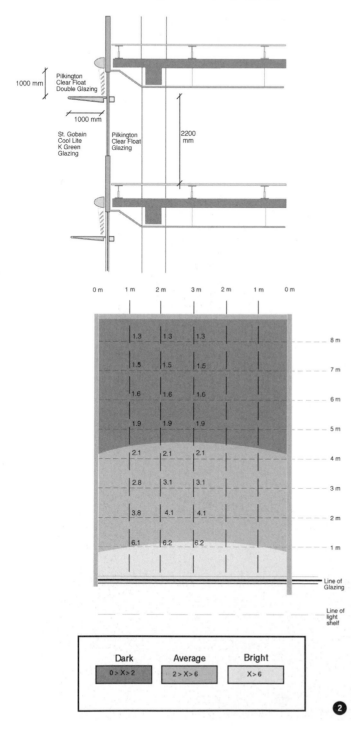

Partnership, KPF, Troughton McAslan, Harper Mackay and others.

The founding partners maintain a close personal involvement with all projects — providing or laying down the initial concepts/diagrams (often hand drawn). Each project is administered by an overall project engineer with separate engineering management for structures, services and landscape groups. Staff are allocated to projects depending on programme on a monthly basis using a central resource allocation sheet. The practice consists of 30–35 staff, 25% of whom are engineers and 25% architects/landscape architects. There is a requirement for technicians.

Battle McCarthy IT strategy involves seeking to meet the target of 'every project engineer having a PC and every architect a Mac'. This includes raising computer fluency for the practice. There is no IT/CAD manager although staff are employed in order to develop specialism, for example Piers Heath (mechanical engineer specializing in simulation) and Chris Hillsden (building analyst specializes in the use of computational fluid dynamics models and thermal analysis modelling on a day-to-day basis).

IT applications for text manipulation databases and specification writing

The practice uses Microsoft Office Professional, i.e. Word for word processing, Excel for calculations (especially climatic data representation), graphics, results of analysis. Schematics are often scanned in. The firm uses DTP packages — QuarkXpress for standard templates for faxes, 'Title Blocks for drawings', etc., Photoshop, Freehand, Stratavision and Publisher. They use NES specification on disks and hard copies of NBS specification.

The practice also uses specialist software — Staad for structural engineering work, Tas for thermal analysis, Phoenics for computational fluid dynamics (CFD) all of which are very visual tools. Typically, a project process may often start with building the thermal model before using the CFD model.

IT applications for CAD

The architects in Battle McCarthy use MicroStation and MiniCAD while the engineers use AutoCAD. The practice also uses ClarisCAD and Modelshop.

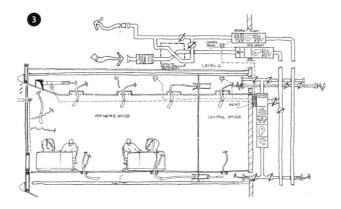

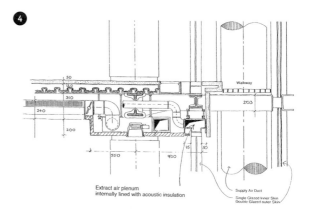

The practice uses an ethernet networked system of desktop and laptop computer terminals running a variety of software backed up by a full colour A3/A4 printing and A0/A1 plotting on in-house hardware. The practice has about seven PCs (two 486s, two Pentium, three Pentiums 2) and 15 Apple PowerMacs.

IT applications for telecommunications and networks

Battle McCarthy originally set up a Mac network then extended it to a half PC network by providing a Mac link. Machines are purchased according to project needs to respond to new software capabilities and to balance income and demand. The practice went through a period of rapid growth and was able to purchase a significant amount of hardware/software in order to set up the BMES workshop.

The building manager also has the administrative task of maintaining the network although capable staff often provide help according to knowledge and experience and the nature of problems.

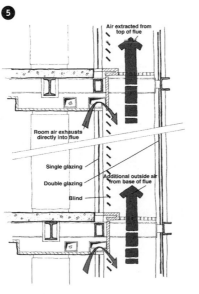

Extract air plenum
internally lined with acoustic insulation

Supply Air Duct
Single Glazed Inner Skin
Double Glazed outer Skin

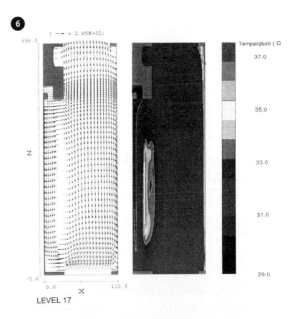

Temperature (°C)

LEVEL 17

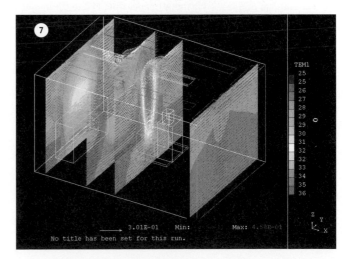

Air extracted from top of flue

Room air exhausts directly into flue

Single glazing

Double glazing

Additional outside air from base of flue

Blind

IT training

Training is provided externally as and when required by specialists. The policy is to employ staff with ability and experience to use certain specific software/packages. In-house training through experienced users is important to supplement external training and support from hardware/software vendors. Specialist software vendors have developed a relationship with Battle McCarthy which allows further development of software through suggestions and writing specific subroutines.

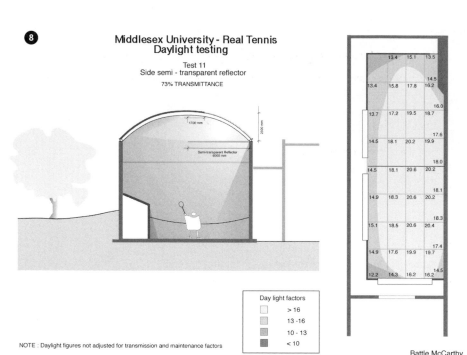

8

**Middlesex University - Real Tennis
Daylight testing**

Test 11
Side semi - transparent reflector
73% TRANSMITTANCE

Day light factors
> 16
13 -16
10 - 13
< 10

NOTE : Daylight figures not adjusted for transmission and maintenance factors

Battle McCarthy

Selected projects

Sime Darby Headquarters, KL, Malaysia, (Hijjas Kasturi Architects)

Services Engineering Building designed with a high efficiency facade utilizing light shelves and low energy displacement air-conditioning system.

1. East and west facades (a) together with north and south facades (b) as freehand sketches. This option yielded 35 W/m² based on model input data
    *   700 mm of opaque (spandrel) panel from FFL to sill height
    *   800 mm of clear double-glazing, above 1700 mm of St Gobain KS477 Cool Light Green
    *   750 mm deep light shelf at 2400 mm height and external fixed shading in front of clear glazing
    *   no internal blinds
    *   2.7 m FFL to U/S of flat ceiling

2. North and south facade analysis — a number of options are tested to optimize thermal performance whilst maintaining a consistency with daylight optimizations
3. An air distribution schematic — roof fresh air plant and local zone mixing AHUs

4. Typical floor bulkhead detail (computer output)

Martini Tower, Brussels, Centre International Rogier, (KPF Architects)

Building Services design for refurbishment of a prominent 30-storey tower, including offices and residential accommodation. In this case the strategy involves the innovative application of a double skin and thermal flue, and an integrated wind and turbine.

5. External wall section showing thermal flue operation
6. CFD thermal analysis on external wall section office/double skin interface, summer, velocity vectors (m/s)
7. CFD thermal analysis office/double skin interface, summer temperature field (°C)

House of Representatives, Nicosia, Cyprus, (KPF Architects)

For drawings and illustrations refer to KPF case study.

Middlesex's University Project

Real Tennis Daylight Testing.

8. Plan and section of selected test 11

BDP Ltd, Architects, Engineers and Surveyors | 5

Background

The origins of the firm can be traced to its founding by George Grenfell Baines in Preston in 1937 on the belief about the organization of design

that it should be on a fully equal multi-disciplinary basis, and that the staff should have a share in the profitability of the firm and therefore a stake in its performance.

This legacy has developed into the BDP values stressing the provision of both design and service

- to work in the spirit of partnership both within the firm and towards all those with whom they collaborate
- to act responsibly to clients, users, the community and BDP people in respecting the environment
- to act ethically at all times.

The present form of BDP as a multi-professional practice of architects, engineers and surveyors dates back to 1961 and since then it has grown to be one of the largest of its kind in Europe providing consultancy services for the design of buildings and urban environments. The firm's goal is 'to be Europe's leading building design practice delivering excellence in design and service through partnerships' aiming

- to create fine work, appreciated by clients, users, public and peers
- to advance the art of creating the built environment giving even better service and added value and perfecting multi-discipline methods to do so

- to sustain and develop a network of offices serving regional and international clients
- to earn sufficient margin to invest in the continuous improvement of people and process
- to be a continuing practice transcending individuals.

The practice has three broad service groups

- (a) *architectural*: architecture, interior product and graphic design, landscape architecture, town planning and urban design
- (b) *engineering*: mechanical and electrical engineering, lighting, acoustics, energy, civil and structural engineering

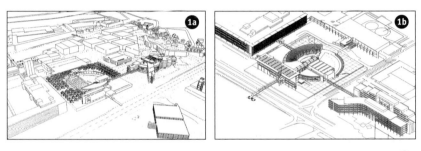

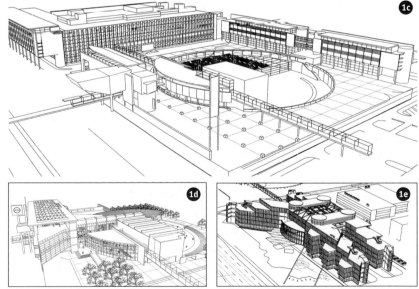

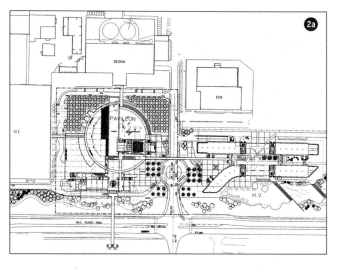

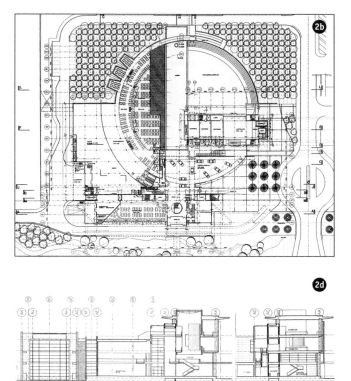

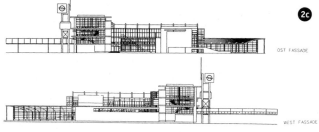

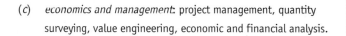

(c)   *economics and management*: project management, quantity surveying, value engineering, economic and financial analysis.

The firm has six offices in Britain and Ireland and a developing network of offices and alliances working worldwide under the collective name of BDP International SC and embracing 1100 people all reflected in one corporate identity — the BDP mark. Each of the offices operates as a separate autonomous organization, under a group of its own partners who set targets, market services, supervise the design and management of projects including recruitment and firing of staff. The central organization contributes some services such as Quality Assurance and IT to all offices under the chairman who is responsible for presenting the 'BDP brand' to the world and the chief executive in control of major management

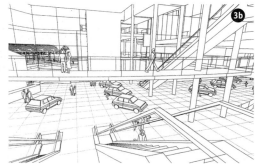

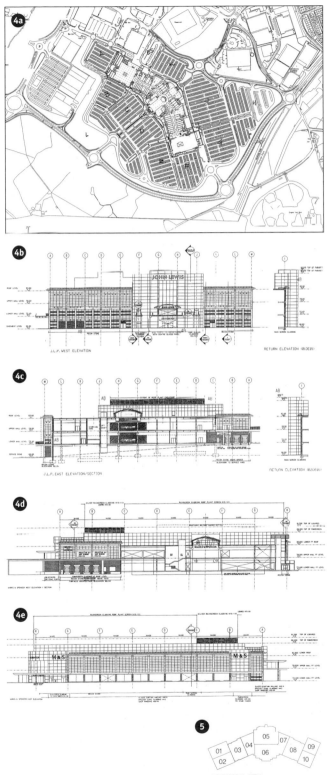

decisions for the whole group of offices. Both the chairman and the chief executive are elected by all the partners for a two-year term.

BDP enjoys a great diversity of activity corresponding to the geographical range of work following initiatives across Europe. The firm's success, apparent in the diverse portfolio of work, is based upon a profound knowledge of building types, design techniques and construction. The focus is on multi-specialist competence with a philosophy that design and performance are best when the participants work collectively together in multi-professional teams rather than setting out to enable a particular design approach or the cult of the prima donna to dominate. BDP points out that the firm pioneered the concept of multi-disciplinary working in the UK, allowing collaboration with all those involved and with client participation to fulfil their aspirations while providing single-point responsibility and a single line of communication which offers benefits of

6b

organization and managing the design process. Since the early days systematic record-keeping of numbers of staff, structure of offices and locations, and the breakdown of output according to the range of disciplines have been important characteristics which have led BDP to develop well thought out management systems.

BDP projects are carried out with primary objectives of completion to programme, working to budget and achieving integrity, quality, appropriateness and added value. Each project is led by a job director who carries responsibility for the BDP service. In complex and large projects a design director is often appointed alongside the job director. Design team members are selected for their appropriate skills and experience and to embrace effective team response.

Paul Davies, IT manager (a structural engineer by training) heads the BDP IT group of ten staff with IT support groups (varying in size from six full-time to two part-time) in various BDP offices. The IT group covers corporate database, corporate purchasing and operating systems, WAN, LAN, general office activities, and CAD standards (quality procedures, naming conventions, etc.). An IT managers' group meets about 2–3 times a year. There is also a CAD standards focus group. IT for BDP has been driven through the organization and especially through requirements of large projects.

BDP involvement in IT began in 1979/80 driven by R & D requirements for CAD systems. At the time BDP looked at Intergraph, GDS, etc. but decided to purchase a core CAD engine (Eagle) which was developed providing an architectural 'front end' Acropolis (with 3D licence). This capability allowed BDP to offer a space-planning service in which the space-analysis program guides the client from a space analysis of its current offices and requirements through ideal

6c

space allocations to specific proposals — all quantifiable, all proven, all highly attractive to a wavering client and in terms of density levels per person, space allocation for each person, flexibility.

Difficulties in use and operation of Acropolis led to development of 2D capability on top of the core CAD engine. This was used on a number of projects for five years. In 1989, because of costs in carrying out software developments, BDP sold its CAD rights and decided to look at industrial packages on PC (AutoCAD, MicroStation, etc.) culminating in the decision to adopt these for BDP on to PCs on a network. It is now policy to use CAD from day one of a project and CAD is used on almost 100% of all projects.

BDP expenditure on IT is approximately £1 million a year involving a continual programme of replacing 286s and 386s. There is an 86% distribution of PCs, 75% of which are Pentiums. An IT strategy is to reduce paper in the office by using more digital exchange. BDP IT strategy is to move to NT servers within a year based on the policies of integration at the data level and an ability to effectively manage the information flow rather than simply sharing applications. This means developing suitable work procedures both contractual (e.g. defining digital, plot lines, etc.) and non-contractual.

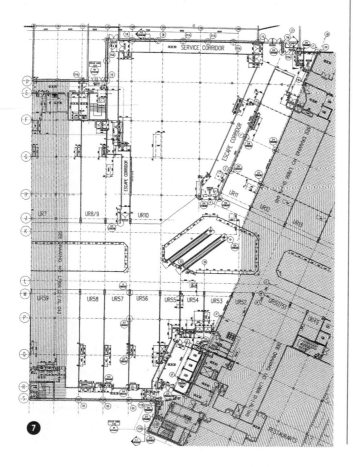

❼

Defined contractual arrangements allow good working relationships with organizations such as Ove Arup & Partners.

IT applications for text manipulation, databases and specification writing

BDP use Microsoft Office after adopting the suite of programs as standard package for all the PCs 'to provide consistency of output whether in Word processed documents, Excel spreadsheets (e.g. drawing schedules) or Access databases'. Microsoft is the standard platform on Windows or Windows NT exploiting the advantages of object link embedding facilities of Windows.

There is a separate graphics department which uses Apple Macs integrated on the network and desktop publishing software such QuarkXpress and Photoshop for magazines and signage with Postscript Printing and high end colour printers. BDP use Specification Manager together with the NBS on disk using Word. NES is used by engineers.

IT applications for CAD

Through CAD work and associated procedures BDP aims to provide quick co-ordination among professions enhanced by application of CAD — compatible technical software. BDP operate AutoCAD (about 150 copies) used primarily by engineers and MicroStation with MicroStation PowerDraft and MicroStation MasterPiece (200 copies supplied by Cadhouse Systems Ltd of Manchester) used mostly by architects. BDP is also a subscriber to the Bentley SELECT program.

Archibus software is used as an electronic tagging tool and linked with AutoCAD can handle tags as a database entry to aid client facilities management as well as the mechanical engineers for services such as air conditioning. The software allows asset tracking and recording of user requirements (room-by-room and department-by-department) and the setting up of a relational database by the interior design team.

Mechanical and electrical engineers also use Hevacomp, AutoPlant, Flovent Product and Ametec. Structural engineers use Integer Suite and FastTrack while civil engineers use Moss software. Cost consultants use software to suit the various disciplines, for example CATO for estimating; Wessex, an expert simulation system, and LC, a cost management package.

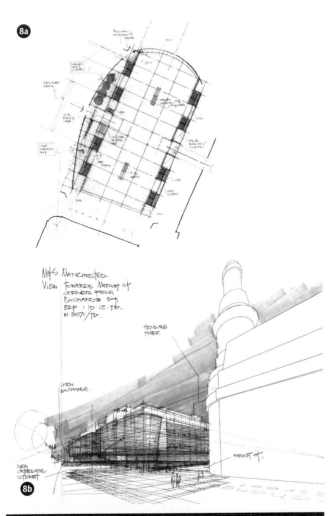

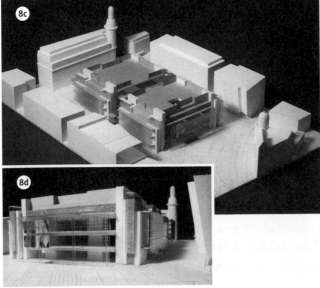

The process in BDP recognizes the three domains of operation: the private domain, the project team domain (shared interface) and the public domain. In the first and second domains the architect works mainly in 2D with a clear separation of 3D work including visualization, animation and publishing for client briefing. Site plans constitute the third domain. The graphics department is heavily involved in this latter area with an emphasis on need and clarity — 'why one does something'.

IT applications for telecommunications and networks

The situation has changed from the time the PC Novell network was mainly used for administration purposes and by secretaries. Now the network is integrated with all aspects of BDP activities.

Each office has a local area network (LAN) linking desktop PCs. The LANs are connected to form a wider area network (WAN). BDP has had lease lines for about two and half years from Manchester directly to London, Glasgow, Sheffield and Belfast, indirectly to Dublin via the Belfast offices. One of the most likely future developments is an increase in bandwidth.

BDP use a number of designations: File Transfer (including E-mail) and File Permissions (J: drive for administration; L: drive for CAD and Home Area; K: shared drive for named persons). Back-ups are carried out every night and quality issues are taken seriously to safeguard BDP's legal position, e.g. 'issuing information without title block'.

IT training

BDP policy on training is set within a well-defined context of career progression for individuals according to five, seven-year career stages (1–running yourself for the newly qualified, 2–running jobs, 3–running groups of jobs, 4–running one of the offices and 5–running out). On IT the policy is not to employ CAD draftsman or engineering technicians but to encourage use of IT applications by staff, e.g. use of AutoCAD LT on laptops and desktops. Staff are also routinely sent on three day CAD courses for training.

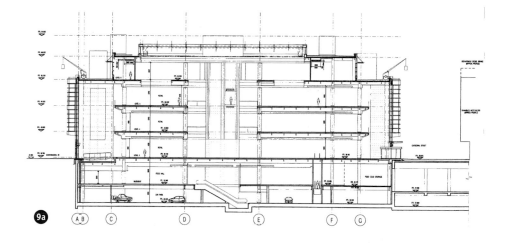

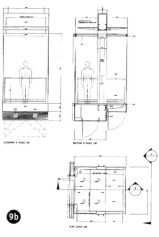

Other issues

Paul Davies, BDP IT manager questions

Why the automotive and aircraft industries have had a great deal of success in applying IT while the construction industry is finding it difficult to do so. The problems of culture and risk need to be overcome.

Another issue of importance to BDP is the idea of using the same data model to provide the ability to move seamlessly between disciplines linking applications ensuring inter-operability. Digital exchange is an area on which the firm has spent a lot of development time, evident in five to six methods of being able to exchange data externally using routers, analogue and digital exchange telephone lines.

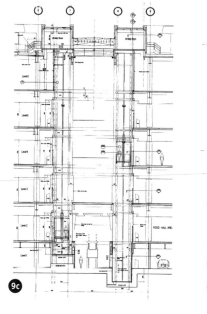

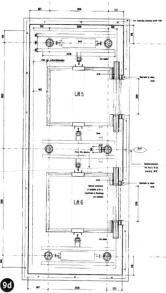

Selected projects

The Royal Albert Hall refurbishment, London

This project is important in its use of the Document Management System adopted for the project by the project team.

Adam Opel AG Headquarters and Communication Centre, Germany

This project gave BDP the opportunity from the outset to use CAD and IT as an integral part of the design process working between both paper and screen. An NT exchange server was set up for the project to allow exchanging documents, sending files and mailing messages between BDP and Germany.

1. (a–e) Site terrain and many propositions were modelled in 2D and 3D using first Acropolis and then MicroStation V4–95. The images have been assembled in a booklet and video to explain the project and how it works to future occupants
2. (a–d) Plans, sections and elevations
3. (a–b) Interior views generated using computers

The Cribbs Causeway Shopping Centre

This project was for clients Prudential and JT Bayliss & Company. The Cribbs project indicates inspiration and enthusiasm for the electronic management system by the client and within the project team, although it cannot be relied upon to guarantee the desired smooth digital data exchange.

4. (a–e) Plans, sections and elevations show the design of 700 000 square feet (63 000 m^2) of retail space on a 108 acre site near Bristol, UK, and parking for 7000 cars as well as a bus station and a bus park. The centre is dumb-bell shape, angled in the middle, and constructed using a glass roof with a solar shading system
5. The location key plan is one way in which the complex scheme was broken down and made more manageable
6. (a–c) Many 3D images allowed visualization of the individual shops by tenants and the client allowing pre-sale of the units
7. CAD helped the integration of structure and services, e.g. vertical location of service cores

The solar shading system gives 50% solar shading throughout the day and provides a surface with 50% reflectivity against which to bounce interior light at night. BDP maintains that a major feature of the design is the energy-efficient displacement ventilation system, which brings in low-velocity treated air at ground levels, which then rises slowly through the building.

Marks & Spencer Store, Manchester

The store has an ultimate trading potential of 22 500 m^2. Marks & Spencer is an international organization that began life in 1884 as a stall in Leeds Market growing into 600 stores worldwide and including US-based Brooks Brothers. The Marks & Spencer brands of clothing, food and housewares have become household names throughout Europe. The store design uses a central atrium to house the elevators and planar glazing to capitalize on the excellent adjacent views and allow customers easier orientation across such a wide store.

BDP were appointed in the late autumn of 1996 as the architects for the new flagship store in the centre of Manchester. Sketch options were developed in the early stages, together with physical site massing models, and outline design proposals were presented in January 1997. A detailed computer model of the city centre was also completed at this time and early street level views were set up to assist massing and detailed design development.

8. BDP utilized hand-drawn sketch proposals and physical models to explore the store's design. Once the fundamental design principles were established the two-dimensional kernel of information was transcribed into CAD format, allowing the high level of design development to be accommodated easily. The client stipulated that all consultants were to utilize compatible ISDN cards for data exchange. The exchange of data between members of the design team was overseen by BDP's project CAD co-ordinator, who was instrumental in the choice of MicroStation as BDP's project software. The software's reference file system, together with its seamless ray-tracing visualization capabilities, were key issues in this choice.
9. 2D plan, sections, elevations, lift package details, etc. were used as the basis for contract documentation generally. A major part of the modelling effort on the 2D computer models related to the control of temporary models used to explore client-instigated design options, often exploring several options simultaneously, while maintaining a core set of models relating to approved design proposals. This control was fundamental in allowing construction information to be developed for the structure and building envelope while fitting-out options were still under active consideration. The office utilized 'Specman' to manage the work package specifications, and this system allowed BDP to draw upon and add to the NBS system.

BDP chose to generate an in-house bespoke 3D computer model. This was used to explore a number of key views around and inside the store. The images were used to present the proposals to the client, in discussion with the planning department, and were also used to convey design intent within various work packages. The model was used to generate an animation of the building's exterior, which was presented to the client and used to evaluate signage and marketing issues. The technique allowed the design to be expediently fine tuned.

Buro Happold, Consulting Engineers | 6

Background

Buro Happold was established in Bath in 1976 as a multi-disciplinary practice of consulting engineers in civil and structural engineering, building services and environmental engineering, and infrastructure and traffic engineering. Five founding partners include Ted Happold and John Morrison. The practice has grown to be worldwide with offices in, among other places, London, Leeds, Hong Kong, New York. Areas of special expertise include: (FEDRA) fire engineering design and risk assessment; long span and lightweight structures; lighting design; MEQS and electrical services quantity surveying and cost control; BHS quantity surveying and construction cost consultancy; window cladding technology; (ECFM) engineering consultancy for facilities management; engineering geology; BHEA environmental assessment and BHPM project management.

The practice's guiding principle is that engineering considerations influence planning and architec-ture and that good designs result only when there is a genuine harmony between the artistic and the practical.

The practice objectives include securing value for money and completion of projects within predetermined time and cost parameters. The practice's underlying ethos is the use of technological innovation wherever it is appropriate practising as separate design groups under ultimate supervision of a Board. Each group is led by a partner who is responsible to the Board and within each group individual group managers undertake responsibility for the execution of projects while responsibility for the direction remains with the initiating partner.

All projects are subject to Buro Happold quality assurance procedures and cover a wide range of work including: site development and infrastructure; development of land for environmental, recreational and leisure purposes; schools and university buildings; sports and stadia; hotels and residential buildings; religious and historic buildings; museums and galleries; office buildings; complexes and commercial developments; theatres; Law Courts and other public buildings; and car parks and open spaces. Customized solutions using CAD throughout any project's lifespan are offered by the firm. This includes development of CAD standards and procedures; CAD data exchange advice to project teams, customization of CAD software, macros, routines, etc. and membership of external bodies.

The practice seeks to use the full potential of computing power in design, documentation and specification and computer modelling supporting all areas of work. This means including the more technical specialist areas such as finite element and dynamic relaxation structural analysis, thermal analysis, lighting analysis, the establishment and analysis of site profiles using ground and highway modelling, fabric and cable net patterning and dimensioning.

Buro Happold IT Group is led by IT director Nick Nelson (an associate/senior engineer). IT policy is to be ahead of everyone else, researching the best technology and implementing findings, while providing appropriate infrastructure. For example the London office has an IT budget of £75 000 per year for software and hardware expenditure (approximately £200 000 total for Buro Happold UK wide). Nigel Davis is the CAD manager in Bath. Stuart Small, with Unicorn technical training background, is a consultant employed to provide

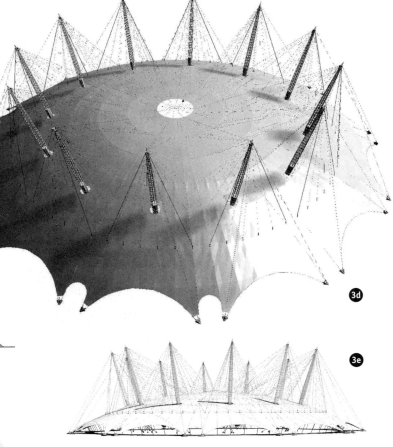

3d

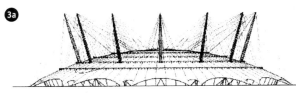

3a

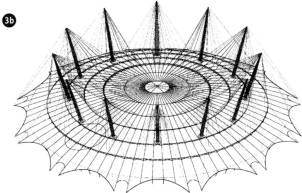

3b

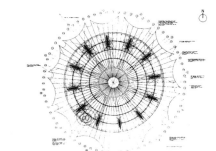

3c

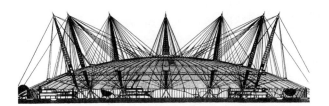

3e

IT management on the premises five days a week. He has been with the London practice for a year. As IT manager for the London office he is generally concerned with the multi-functionality of software packages, for example improving the way AutoCAD interfaces with Microsoft Word.

The firm has a policy of continual updating of both software and technical resources supplemented by specialist programs written in-house — buy new and sell the old approach. Links between offices and between those of clients/professional associations enable instantaneous transmission of design and closer integration at all stages of projects.

IT applications for text manipulation, databases and specification writing

Buro Happold use Microsoft Office Professional, i.e. Microsoft Word for word processing; Microsoft Excel for spreadsheets and for drawing issues; First Class Client for E-mail; QuarkXpress and

Photodelux for scheme reports.

Other software includes: Winsbeck for specification writing; Hevacomp for Mechanical and Electrical, e.g. sizing of ducts, cables; Symap for electrical; Strap for structural analysis (the practice is considering buying analysis software for this purpose) and Cellbeam. Accounts are done separately from the Bath office. Timesheets are done manually and then passed on to the accounts section in Bath.

3f

IT applications for CAD

CAD is used to form the basis of work within the firm and is implemented to complement the engineers' designs and to enhance the project lifecycle. Structural analysis, building services and drafting are all computer-based. The advantages have proven invaluable in both increasing efficiency and the exchange of data.[1]

Buro Happold use the complete CAD software of AutoCAD. The London practice has twelve CAD workstations of which two are Pentium Pros, two PowerMacs (to manipulate in-coming files), one 486 PC (for backup and for manipulating old files) accessed via 60–70 terminals with cross-platform compatibility. Of the 50 staff in the London office 80% are engineers (of whom 40% are computer fluent) and 20% are support (including five CAD technicians, formerly CAD drafters).

The Buro Happold process involves working fast-track, starting with sketches and then proceeding to work by draftsmen to allow sharing and cross-referencing. In general there are two main aspects

(*a*)   internal facades engineering (with director John Morrison having a unique reputation within the UK and worldwide)

(*b*)   structural analysis.

The central modelling department offers 3D CAD model building and 3D co-ordination covering 3D modelling, geometric models (layout and co-ordination work to ease production of 2D plans); production of images and visualization for presentations, schemes and displays; animations and flythroughs, and research and development.[1]

Nigel Davis provides examples of this work in the selected projects.

IT applications for telecommunications and networks

The London practice has a Novell netware 3, 12, 20 Giga byte server P166 Pro, hub netware (four for the 70 terminals) using Windows 95 platform. Rationalization is essential if a move to Windows NT is to be made. There are, however, no immediate plans to do so.

IT training

There is in-house CAD support and training which includes the CAD help line. Stuart Small is responsible for training often organizing lunch time seminars and training sessions for basic things. He uses the web for general keeping abreast of developments and for

(4a)

224 bedroom five-star hotel, residential accommodation, a retail mall, banqueting facilities, underground car parking, landscaping and infrastructure. The 250 m high office tower is square in plan, soaring to a tapered point in one smooth, giant arc, and is clad in silver anodized, aluminium panel. A compact central core contains all the main circulation services; the floors span from core to perimeter columns. Giant K braces set at intervals tie the corner columns together, while defining clear spaces, which act as observation decks, and allow service air intake. The 100 m long banqueting hall lies beneath the main plaza. The tower was designed to hold up to 2400 people in comfort within a long-span column free arched structure. A series of inclined reinforced concrete arches, eight pairs in all, define the roof structure and support the roof slab. These arches converge with one another to create a 'petal' effect, forming a common straight segment, or buttress, to the springing points. Below the hall floor slab, tie beams with post-stressed tendons restrain the buttresses from spreading while a reinforced concrete tie slab between each arch pair acts as a diaphragm, bracing the slender arches. Above the banqueting hall, and placed on either side of the lushly planted plaza, stand hotel and apartment buildings. These buildings will be constructed of pre-cast concrete, local limestone and timber, brought in a multi-layered facade solution that will provide maximum control for internal environments. The retail building is over 250 m long and provides over 34 000 m² of retail space. The building, constructed of long-span, post-tensioned concrete, has been carefully designed to high levels of environmental comfort and fire safety.

The project was a milestone for CAD co-ordination within Buro Happold. The project was built as a 3D model in conjunction with the architects (Sir Norman Foster & Partners) in order to check the cladding and finishes and their relationship to the structure, as well as to help visualize the sometimes complex 3D geometry of the buildings. The architectural model was built as a 'hollow' cladding

support. He is in the process of developing an IT standards manual with a section for the employee's induction manual.

Looking at current IT issues, the year 2000 is unimportant for the practice because they are geared up for it. Another issue is that of security. Despite backing up everyday and weekly and use of waivers on all electronically transferred data there are real concerns which need to be addressed on the wider scale.

Selected projects
These show the nature of development of IT applications within the practice.

Al Faisaliah, Riyadh, Saudi Arabia
This project for King Faisal Foundation is a joint design venture with Architects Sir Norman Foster and Partners. This mixed development is set around an open plaza and features a landmark office tower,

frame, to the necessary criteria and dimensional constraints. The structural frame was then designed to fit within the cladding. Any problems could be identified and flagged very easily. The tower model was used specifically to produce the preliminary 2D floor plans. Sections were cut at each floor to provide layouts of the tapering columns and floor plates, avoiding lengthy calculations of geometry.

1. Macrostructure from the fully co-ordinated 3D building model. Due to the geometrical complexity of the tower a model was built from which sections were taken to provide backgrounds for standard CAD drawings. In this way, the geometry and setting-out could be analysed, co-ordinated and checked at any point

2. 2D plans generated directly from 3D model reflect dynamic project file-sharing with all disciplines and developed project CAD standards for naming conventions, files issues and data storage

Millennium Dome, Greenwich, London

For this project Buro Happold acted as structural, building services engineers and project management.

The practice is part of the engineering team (with architects Richard Rogers Partnership) brought together to build the dome which will house the National Millennium Exhibition and celebrations. The dome is thought to be the world's largest domed structure — 50 m high round structure with an overall span of 320 m — dominating the end of the Greenwich peninsula.

According to estimates, over 40 000 people will be in the building at any known time, with 12 million planned to pass through the exhibition in a year. Therefore, careful consideration of the enormous use the building will be put to is vital; the building will use the same amount of fresh water as up to 5500 average households, and at any one time, will consume the same amount of energy as 30 000, 100 watt light bulbs. The pavillion contains an exhibition geared around zoned themes and has a translucent tented roof suspended from cables running from 12 100 m high masts. The glass sides and white fabric roof will dominate the London skyline by day, and allow it to glow spectacularly by night. There is a complete CAD data co-ordination for the whole project and full project CAD standards across all disciplines. Virtual project offices were implemented to dynamically link each design team to site offices.

The consultants and the two construction management practices (Sir Robert McAlpine and John Laing Construction) adopted the MicroStation CAD system for the project and established basic common ground, e.g. file-naming conventions and directory structures, without trying to impose complete uniformity. To cope with the project requirements Buro Happold increased the number of MicroStation terminals from 16 to 26. At first the intention was to use MicroStation 95 for the whole job but then a change was made to the subscription-based SELECT version. The VRML output that this offers has been

4b

4c

enormously helpful in design validation, e.g. in resolving the complex geometries at the point where the dome is penetrated by the Blackwall Tunnel vent.

The use of 3D modelling on the Millennium Dome had proven to be an invaluable tool. The dome is not a standard object that lends itself to simple 2D drafting. In this project a complete model of the dome structure was built for co-ordination and analysis. The level of detail was literally to every bolt; additional buildings and structures have been used for both analysis and design of the structures and have been added to the 'virtual' site as the project progresses. The 3D model contains an unprecedented amount of information and has been at the centre of design development. The live design data is stored at the Greenwich site office, with both the Richard Rogers Partnership and Buro Happold offices able to access information via ISDN links. Approved drawings are signed off and passed to McAlpine Laing, so that anyone logging onto the server can access them knowing that they are the most up-to-date. The fabric structure was the responsibility of Buro Happold's special structures division, which uses its own proprietary UNIX-based system for complex geometrics of fabric structures. At an early stage, therefore, data from this was output via DXF and incorporated into the live 3D model at the central Greenwich office. The fabric model had been used for both analysis and design of the structure, and has been particularly useful in investigating the effect of wind and snow using Buro Happold's TENSYL software. Also, owing to the exact co-ordinate placement taken from an Ordinance Survey grid of the Dome model, site co-ordination has proven very easy, both with the 3D model and the related 2D plans. Animations and simple virtual worlds have been used to study certain aspects of geometric clashes. Without the 3D model these checks would have been slow and laborious.

3. (a–f) 3D models for the scheme and for working drawing production facilitated by expanded data translation to provide complete integration with all disciplines CAD packages

Harbourside Development, Bristol

This project is for the Harbourside Centre in which Buro Happold are structural, geotechnical and civil engineers. The £54.8 million arts centre complex by architects Benisch, Behnisch & Partner

Architeckten has two auditoriums (one to seat 2300 people, the second a dance theatre to seat 400), back stage areas including rehearsal spaces, restaurant, ticket office, and extensive foyer spaces cantilevering over Bristol docks with piles within the docks. Important aspects covered

- complex rethink of CAD data production to allow for structural complexity of the project
- object-orientated design approach adopted for building elements
- database links to CAD data to provide on-line materials information.

4. The Harbourside Centre, Bristol's new centre for performing arts, is another giant leap in CAD data production for Buro Happold. The project's 'random' design and asymmetrical aspect has forced the CAD department to develop new methods. The whole of the structure will be produced as an object-based model. Intelligent attributes will be attached to model elements, allowing full measuring and geometric analysis to be automated. The processes involved have been adapted from other engineering disciplines' standard methods, particularly industries such as automobile manufacturing

BAA plc Framework Projects

For this project Buro Happold provided structural engineering services as a partnership with a client who is fully committed to adopting a process of continuous improvement. The framework agreement is regarded as 'an exciting and stimulating opportunity to explore new, efficient ways of streamlining the design process to bring value to all projects. Useful comments concern

- analysis of design and data production methods to streamline and improve efficiency
- move towards true 'collaborative engineering' and data sharing
- use of Internet and project intranet for access to CAD data
- integration of electronic data management systems for control of digital data.

References

1. Buro Happold practice statement.

Chris Wilkinson Architects | 7

Background

Before setting up practice in central London in 1983 with partner James Eyre and 13 years after graduation at Regent Street Polytechnic (now University of Westminster), Chris Wilkinson worked formerly with Lasdun's office in the 1970s, then Norman Foster, Michael Hopkins and Richard Rogers & Partners. Chris Wilkinson has also written about supersheds. A supershed is defined as

a kind of architecture which is not formal, decorated or mannered, but which derives its aesthetic from a simplicity of form and construction and a clear expression of its purpose and component parts.[1]

At the beginning of the 1990s the practice moved to the Clerkenwell district (three large former classrooms with walls painted white, window sills set determinedly high in the walls) and into its current offices in Old Street. The offices house 29 staff, the majority of whom are architects.

The projects undertaken by the firm range from the design of commercial, industrial,

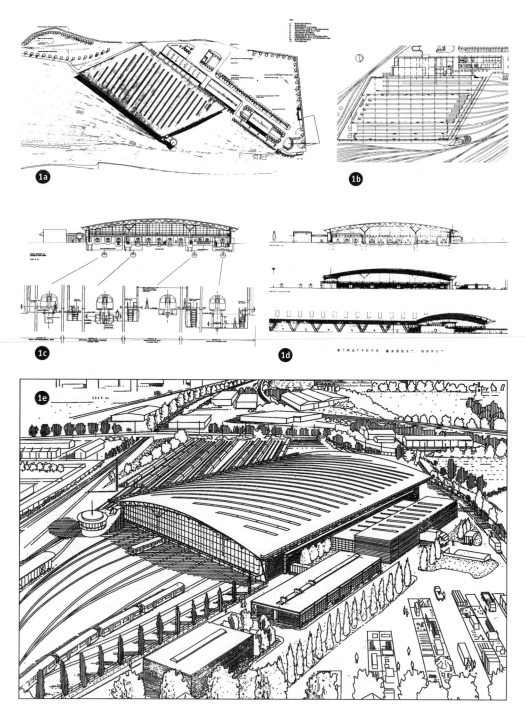

transportation, residential, sport to leisure buildings as well as master planning, interior and component design and contract management. Each project team is led by a project architect. The CAD manager co-ordinates all computer activities. The 1994 Hayes Davidson computer-generated aerial view of Wilkinson's Jubilee Line maintenance depot has now appeared in so many national and international publications that it has become part of the great genetic pool of architectural ideas (eg. Fig. 2 (a-c)).

CWA have invested heavily in CAD systems. Back in 1990 when design work on the Jubilee Line depot began, the firm used to rent CAD equipment by the week with staff working shifts to get value for money out of the machines.

IT applications for text manipulation, databases and specification writing

For standard applications CWA use Microsoft Word versions for Macintosh for text manipulation, letter/memorandum writing, report writing and documents incorporating very simple graphics and diagrams. For specification writing CWA use NBS disk Microsoft Word versions for Macintosh (available for reference purposes only). Microsoft's Excel spreadsheet is used for tabular and calculated information such as drawing issue sheets, telephone lists, area schedules, fee proposals, general cost calculations, etc. which can be

used to create Excel charts, diagrams and pie-charts.

Special applications used include FileMaker Pro, a simple database program, and the accounting system, Access Accounts. CWA also use QuarkXpress for desktop publishing software and for use in complex page layouts and documents with a high proportion of graphics,

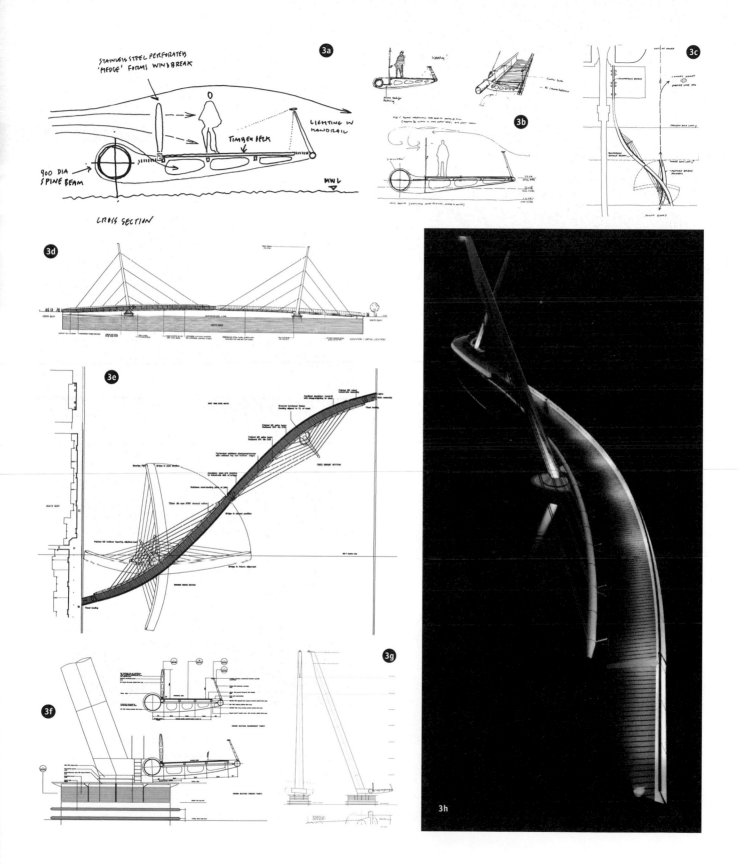

either images or drawings, i.e. special text layouts which cannot be performed in Word, typically fitting text with text flow to non-orthogonal shapes.

IT applications for CAD

The firm uses MicroStation on Apple. Specialist CAD software includes one version of Minicad version 7; Form Z for 3D modelling which is used for the majority of the work especially on bridges and strata, and Studio Pro 2 for rendering.

The processes and activities follow from hand sketches to 2D drafting using MicroStation, then 3D modelling using Form Z and finally rendering using Strat Studio.

IT applications for telecommunications and networks

CWA hardware is expanding but currently includes: Apple Macs (four PowerMacs, three Quadros, a total of eight LC2 and LC3s); ten CAD workstations and desktop publishing; five administration machines and one dedicated E-mail machine. There is an Oce 9400 laser plotter (otherwise they send out for large colour printing) and for four years the ethernet has been connected to a central Apple Mac server.

The range of machines covers: three LCII 4/40 Apple Mac; one II SI 9/40 Apple Mac and three Quadra 650 Apple Mac for administration; two IICI Apple Mac for CAD stations; one PowerMac 8500 Apple for 3D models and CAD; one PowerMac 9500 Apple for 3D models/rendering; two PowerMac

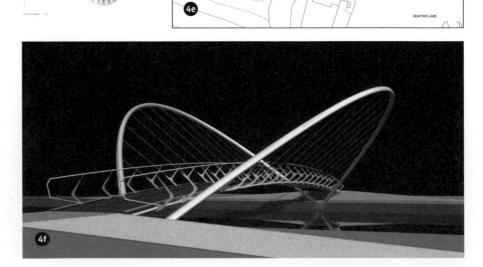

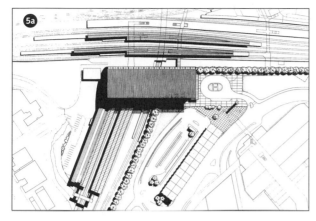

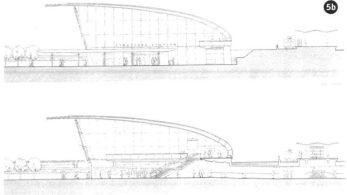

7300 Apple for CAD stations and six PowerMac G3 for CAD and desktop publishing using QuarkXpress. Fileserver is a PowerMac 9650/350.

IT training

CWA use in-house training based on the knowledge of experienced users.

Selected projects

These are used for reference on IT applications by CWA.

Stratford Market Maintenance Depot (1994–6)

An epic structure of 60 m spans with a 100 m of curved roof 190 m long and 36 000 m² roof area allows space for the inspection, maintenance and repairs of 11 complete trains at a time. The project for the client Jubilee Line Extension Project was won in a competition held by London Underground Ltd in May, 1991.

1. (a–e) Plans, section and elevation indicate straightforward 2D computer outputs. The parallelogram shape of the main building is derived from tight site constraints and the track alignment requirements to accommodate the minimum radius and allow straight track within the building. Secure stabling for 33 trains is provided on the eastern part of the site, administered from the control building attached to the north-west corner of the shed. The traction substation, administrative and amenities building, together with stores and workshops, are located in a line along the western side of the shed with direct links and multiple services connections

2. (a–c) 3D computer images of the building enhance visualization of the overall form of the 30 degree 'diagrid' roof structure with 2 m deep arched lattice trusses at 9 m centres crossing to form a space structure. This is supported on an intermediate grid of tree-like columns at 18 m x 40 m centres, which branch out to connect to the node of the diagrid roof, with 'V' columns on the perimeter. Rendered images are by Hayes Davidson.

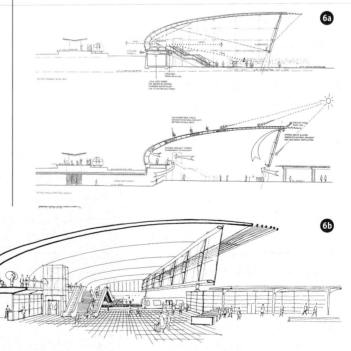

INTERNAL PERSPECTIVE

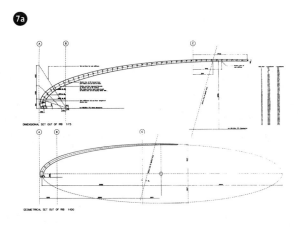

7a

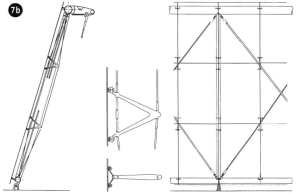

7b

Bridges

CWA have acquired a reputation for designing high profile bridges. This is mostly as a result of winning design competitions for bridges and for 'economical landmarks', including the following

- 1994 London Docklands Development Corporation for a 180 m span, cable-stayed, 'S-shaped' dock-crossing (featuring two raking masts and a sinuous deck curved in both plan and elevation) at Canary Wharf, engineered by Jan Bobrowski

- 1995 the diagonal suspension arch at Hulme on the outskirts of Manchester engineered by Ove Arup & Partners

- 1995 the 'invisible' wire supported bridge for the Science Museum at South Kensington in London by Bryn Bird

- 1995 the daring double-raked arch suspension bridge spanning 30 m between willow-lined grassed banks at Russell Park in Bedford with engineer Jan Bobrowski.

3. (a–h) Sketches, computer analysis and drawings are an integral part of the design and evolution of the concept of South Quay Footbridge, completed in May, 1997. This integration extends to the combination of sculptural geometrical form and the functional requirements. Rendered sculptural computer image is by Hayes Davidson.

4. (a–f) Embankment Renaissance Bridge[2] on the Upper River, Bedford, completed in November, 1997, also shows the design process within the practice beginning with early conceptual sketches, through to 2D computer drawings, structural analysis diagrams and 3D computer images. In this case the computer images representing 'the collaborative application of art and engineering in contrast to Webster's engineering latterly perceived as art' are produced in-house using MicroStation/Strata Vision software.

The outward leaning arches are an architectural device giving the bridge a column-enclosing space, yet open to the sky. This form also makes more of the experience of passing beneath the bridge — as rowing eights do — seeing the structure's 'fifth elevation'. Jan

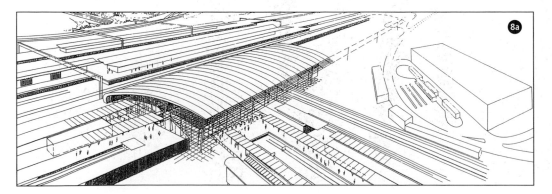

8a

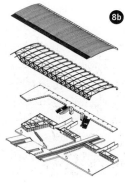

8b

9a

Bobrowski & Partners' computer modelling tested Eyre and Keith Brownlie, the project architects' intuitive structural response, while the architect's CAD models refined the proportions.

Stratford Regional Station Redevelopment (1996–8)

This project for London Underground Ltd Jubilee Line Project; Stratford Development Partnership Ltd and London Borough of Newham comprises 4000 m² of new concourse and subways. The project was won in a competition organized by Railtrack, the station owners.

5. (a–b) Plans, sections and elevations indicate the clear expressive form of the building responding to its context, orientation towards the town and ease of expansion. The curved roof springing from the upper level walkway parallel with the main railway lines, sweeps up to a high glazed wall facing the town centre and the Jubilee Line. The lower part of the curved roof is exposed and glazed on the outside face to provide views through to the mainline platforms beyond

6. (a–b) 2D sketches and layouts are invaluable in design development and analysis of environmental factors, structural and physical constraints of the site. Based upon conventional techniques these complement computer drawings. Computer analysis was also used to verify the design concept and to determine optimum thermal properties for the double skin roof. The roof form allows natural lighting and provides solar energy-assisted ventilation in the main space, via the deep void in the double skin roof through which air is drawn by the 'stack effect' and exhausted at the highest point. The natural ventilation maintains air movement and summer temperatures at comfortable levels, it also provides smoke ventilation should a fire occur in the concourse

9b

7. (a–b) Computer analysis facilitated structural rib setting out and design of details of inclined glass wall showing steel elements

8. (a–b) Aerial view of the building and the exploded drawings enhance visual analysis and evaluation of the construction of the elements and of the circulation patterns

9. (a–b) 3D computer images of the building by Hayes Davidson

Reference

1. Wilkinson C. *Supersheds*. Butterworth Architecture, 1991 and 1996.

2. Wilslocki P. To Ouse, Charm — a new bridge by Chris Wilkinson Architects. *RIBA Journal,* March, 1998, 48–51.

DEGW, Architects, Space Planning and Management Consultancy | 8

Background

The origins of DEGW lie in the formation by Frank Duffy, in 1971, of the London office of JFN Associates. JFN were a leading firm of space planners, with their main office in New York and a European office in Brussels. In 1973 Frank Duffy left JFN to form DEGW Partnership with John Worthington, Peter Eley and Luigi Giffone (who had also worked with JFN in Brussels). Frank Duffy, Peter Eley and John Worthington had studied together at the Architectural Association in London, graduating in 1964. After graduation they gained experience in a range of different offices, spending time in the US on Harkness Fellowships.

The DEGW Partnership was established as an independent London-based architectural and space planning firm. One consistent objective for the firm has been an emphasis on buildings from the user's point of view while concentrating its services in areas that do not depend on new building work, for example interior space planning, strategic appraisal of buildings, brief writing and strategic facilities planning. The client user being more than the individual occupant working in the building but also the wider organization that inhabits the building over its life span.

Geographical development of DEGW occurred in the 1980s, with the establishment of offices in Glasgow, Madrid, Milan and Paris, and in the 1990s Amerfoort, Berlin, New York and Sidney were added. In some cases these are joint ventures with local firms in disciplines of architectural design, furniture and product design (e.g. DEGW T3), behaviour of people in buildings (e.g. Buildings Use Studies), information technology (DEGW ETL), graphics and publishing (e.g. Bulstrode Press). The reputation of the practice as specialists in the design of the Workplace has been enhanced by articles in professional journals, books and published research reports into organiza-tions and their relationships to the buildings they inhabit. This has been instrumental in developing a formal marketing approach. In 1989 the partnership became incorporated.

In 1996 DEGW merged with Twijnstra Gudde, Management Consultants and is now a wholly-owned part of the Twijnstra Gudde Group, a privately owned group of consultancies. TG is the largest independent management consultancy in the Netherlands based on three services of management consultancy, project management and interim management. The group also includes Marketing Improvements, a marketing, research and training consultancy based in the UK who also operate internationally.

Notable landmarks in the development of DEGW during the 1970s were

- design of a headquarters building in Warrington for British Nuclear Fuels
- several design and space planning projects for IBM UK
- several projects for the auction house Christies
- publication in 1976 of *Planning office space*[1] by Frank Duffy, John Worthington and Colin Cave with articles in the *Architectural Journal*, 1973.[2]

Notable landmarks in the development of DEGW during the 1980s were

- briefing for Stockley Park and Broadgate developments
- interior design for Lloyds Bank at Hays Wharf (Office of the Year Award 1990)
- interior design for the 1958 building of Lloyds Corporation

1

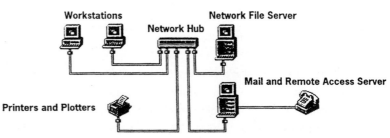

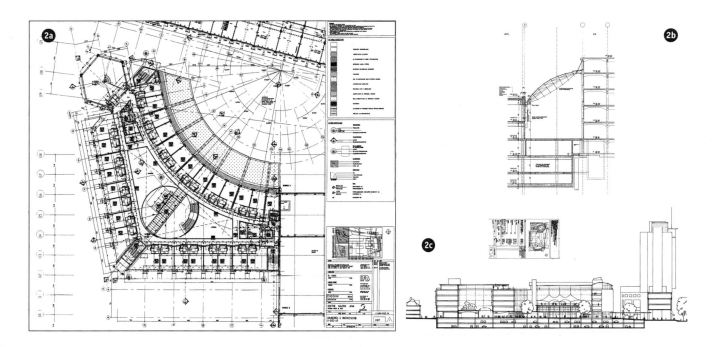

- The ORBIT multi-client studies on the impact of information technology on office buildings (Orbit 1 in 1983–4, and Orbit 2 in the US in 1985)[3,4]

- *The Changing City* published in 1989.[5]

Notable landmarks in the development of DEGW during the 1990s have been

- completion of new pathology laboratories for The Great Ormond Street Hospital

- design and construction of a research and development centre for Olivetti in Bari, Italy

- winning of the design competitions for large industrial and office developments in Wedding, Berlin and Jena, East Germany

- multi-clients studies on the Responsible Workplace, Intelligent Building in Europe, New Environments for Working, and establishment of the Workplace Forum research programme

- new headquarters for the Department of Trade & Industry

- space planning and briefing of main building, Whitehall for MOD

- winning of international competition for Apicorp in Saudi Arabia

- intelligent building in South-East Asia, multi-client study with Ove-Arup and Partners and Northcrofts

- new headquarters for the Boots Company, Nottingham

- interiors and space planning for Grand Metropolitan Plc, London

- programme management in Europe, the Middle East and Africa for Cisco Systems Inc.

In the organization structure DEGW Group Ltd is above local offices, each led by a managing director who runs units. The unit leaders manage the day-to-day management within the units — planning resources, the progress of projects and with the responsibility for the profitability of the unit and cross charging of staff when they work within each other's units. Each unit manages its own paper work on standard formats, developing its own approach to quality control of output. The London office is divided by product, for example building appraisal and strategic facilities planning. The central group unit manages joint affairs, some aspects of publicity and some centralized services to assist day-to-day operations.

DEGW aims to deliver value to clients and end users through design of places that can adapt to the changing needs of the people and processes, and help to enhance their performance and productivity, 'integrating people, processes and places'. DEGW see Information Technology as a powerful catalyst to make a difference in work process and office design. IT is seen as an essential tool enabling them to perform effectively and efficiently throughout the design and production process. It enables internal communication between project team members, externally with other DEGW offices, with clients, other professionals and suppliers.

Important IT staff in DEGW, London are Andrew Harrison, Ian Hall, Ron Harrison (formerly with Rank Xerox) and Tony Thomson (formerly with Hewlett Packard). Colin Cave (chairman 1989–1995) explained that notable events in the development of IT within DEGW were

- adoption of Ceefax in the late 1970s

- purchase of Apple Macs in 1982 and IBM PCs (which had arrived in 1981) following projects with clients such as Digital, IBM. In particular the NOSS Study for IBM 'helped us to see the three levels of network: primary, secondary distribution (vertical) and tertiary (floor)'

- installation of fax by the client in 1983 (Single Body)

- arrival of CAD in 1985 with Andrew Harrison as system manager and Andrew Parson as CAD manager

- arrival of microcomputing and full networking (innovative twisted pair network and first server in 1989) coincided with an office move for the 150 staff to Porters North at King's Cross, London. The opportunity of demonstrating the Workplace strategy and design expertise of the practice was then lost. The recession, 1990–95, meant dramatic cut-backs on staff and an emphasis on technical development with a limited number of staff

- Qmail arrived in 1992

- integrated accounts and professional computing in 1995 produced general reports and monthly management information to business units' directors.

IT applications for text manipulation, databases and specification writing

The London office have installed Microsoft Office which is used to produce high quality proposals, reports and presentations. Word is used for word processing; Excel for spreadsheets and PowerPoint for presentations.

4D Database is used for database applications by twelve trained staff. Project Power software is provided on two workstations and is used by five trained staff.

IT applications for CAD

Desktop workstations are a mixture of Apple Macintoshes and PCs with Apple Macintoshes used primarily as general purpose workstations and for document creation, whereas PCs are used as CAD workstations running 15 copies of AutoCAD by a total of 30 staff. On the CAD network (Fig. 1) all current project information is stored centrally on the network file server and all programmes, such as AutoCAD and Word, are run locally on each station. DEGW have one copy of Archibus.

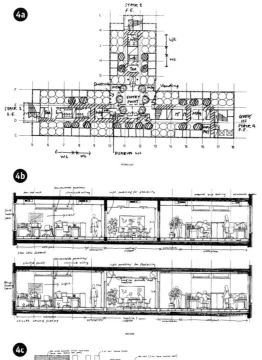

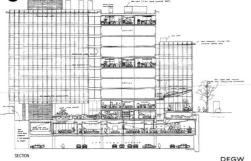

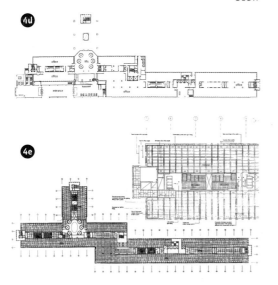

IT applications for telecommunications and networks

Strategically DEGW aim to migrate to PC-based networking for all activities. For this purpose a number of central NT servers have been installed combined with a range of peripheral equipment including A0 plotters, colour and laser printers and high resolution scanners.

The London office has recently upgraded its IT Infrastructure, to incorporate a fully integrated CAT 5 cabling system, in order to provide a flexible working environment supporting new ways of working.

Voice communication within the office is provided by a newly installed SDX Index voice processing system. The system is equipped with full voice mail as well as a cordless phone capability (DECT) for a large proportion of staff regardless of their location within the building. For electronic messaging an E-mail system (QuickMail) has been installed for internal communications with connections to Internet Mail. ISDN links are available and these are used for translation of large files, for example CAD drawings.

IT training

Training is important for DEGW with guidance written in the partnership deed (mid 1980s) emphasizing the responsibility 'to have a concern for professional and personal development of every member of the firm'. In-house support provides a regular programme of IT skills training, aimed at bringing new staff up to the level required within DEGW as well as refreshing and updating the knowledge of the more experienced users. IT support and management is undertaken by an in-house team of specialists. Hardware maintenance is also undertaken in-house.

Issues of concern to the practice include the establishment of a web site. Workplace Forum web site has already been established since January 1998. The aim is to accommodate new kinds of technology, dissolving boundaries between office and home and detailing the latest trends in products and services worldwide.

Selected projects

The projects show aspects of learning from clients as indicated in adoption of IBM PCs in the late 1980s.

Jena, East Germany

This was a large post-industrial regeneration office development project. The masterplan creates a rich mix of uses, including retail and commercial activities, a new university library and public amenity spaces.

2. (a–c) Plan, section and sectional elevations facilitated by CAD modelling show the nature of the geometry involved in the design
3. (a–c) The 3D views were developed using CAD

New Headquarters for the Department of Trade & Industry, London

A £61.4 million refurbishment to provide low-energy working environment.

4. (a–e) The 2D drawings (plans, sections, elevations and detail sections through the typical office floor) show a mixture of freehand sketches and computer drawings for space planning and other design development activities
5. (a–c) Isometric, perspective, external and interior sketches indicate the various 3D outputs

New Headquarters for the Boots Company, Nottingham

A requirement of the project was that all consultants operate compatible CAD systems and the sponsor was keen to be part of a 'smart' way of working that will result in a fully co-ordinated electronic record of the buildings that can be a basis of an integrated management and maintenance programme.

6. (a–f) Following granting of planning permission for refurbishment of the 20 000 m² heritage listed 'D20' building, one of

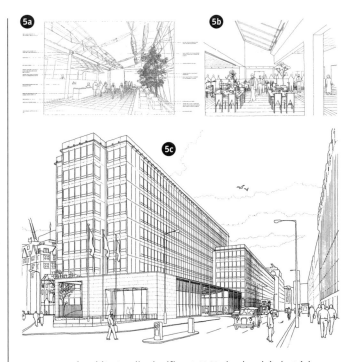

several architecturally significant 1960s landmark industrial buildings in the UK, and for construction of a new 18 000 m² building alongside, DEGW is proceeding with detailed design. Intelligent building principles, identified in continuing research, are being applied to the building design to improve energy efficiency and to maximize the ability to adapt the usage of the

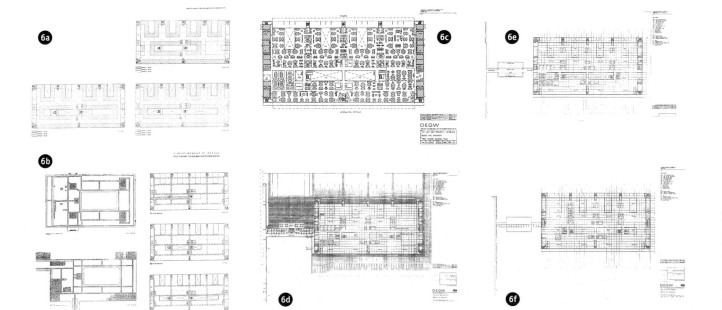

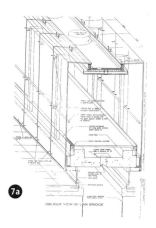

7a

7b

buildings over time. The 2D drawings (plans, sections, elevations and details) show the role played by the computer in space planning activities involving furniture arrangements, circulation strategies, quality of usable space measured against external awareness, etc.

7. (a–d) The 3D illustration (link details and interior views) were facilitated by CAD modelling as the design is driven by user requirements in structured and highly participatory communication process

Re-design of DEGW London offices

The London offices were re-designed in order to support the way the firm works across multiple locations, both in and out of the office — linking people, process and place. The design is based on focusing on the client and on the work which needs to be done. Mobile and cordless technology play a vital and integrating role with

(a) all mobile workers (50% of the population) supported by laptops and cordless telephones and no longer bound to particular desks or locations within the office

(b) project groups, which although more predictable in location can re-arrange themselves as their workload changes because their project files are now centralized and a networked IT platform serves both PC and CAD users.

Crinan Street office layout is based upon the appropriate information for a range of user types — the nomadic worker; manager of multiple teams; independent; team resident; support and visitor. Further definition is in terms of work settings, clear desk policy, technology, booking/reservation system, voice mail and logging out.

References

1. Duffy F., Worthington J. and Cave C. Planning office space. *Architectural Press*, UK, 1976.

2. Duff F., Worthington J., Cave C. and Deakin J. AJ Handbook: Office Building. Section 1 in *Architects Journal*, **157**, No. 18, 1973, 2 May, 1057–1081; Section 2 in *Architects Journal*, **157**, No. 20, 1973, 16 May, 1191–1221; Section 3 in *Architects Journal*, **157**, No. 22, 1973, 30 May, 1325–1342; Section 4 in *Architects Journal*, **157**, No. 24, 1973, 13 June, 1453–1458; Section 4 in *Architects Journal*, **158**, No. 28, 1973, 11 July, 105–111; Section 4 in *Architects Journal*, **158**, No. 30, 1973, 25 July, 219-223.

3. Orbit 1, Building Use Studies, DEGW, Eosys Facilities. Bulstrode Press, UK, 1983–4.

4. Orbit 2, DEGW (with Frank Becker and Gerald Davis), Harbinger, Facilities Research Associates Eleven Contemporary Office Buildings, Rosehaugh Stanhope Developments. *Meeting the Needs of Modern Industry*. Stockley Plc, 1985.

5. DEGW. *The changing city*. Bulstrode Press, 1989.
   Information Technology and buildings. Butler Cox Créer les Espaces de Bureaux.
   Strafor/Nathan. *Operating your business at Stockley Park, Heathrow*. Stockley Plc.
   IBM Space Occupancy Study, IBM.
   London Underground Station Environment Guidelines. London Underground.

7c

7d

EPR, Architects	9

Background

EPR Architects was formed in 1949 by Cecil Elison, architect, who was later joined by Bill Pack and Alan Roberts. The partnership was incorporated in 1988 forming EPR Architects and EPR Design Company to provide a total planning, architectural and design service. The group's expertise covers offices, retailing, industrial and leisure with the areas of specialization being master planning, architecture, space planning, interior design and model making.

The EPR practice consists of 100 members of staff 40% of whom are architects, another 40% technicians and assistants with the remaining 20% administration, under ten directors. Each director leads each project depending on size and complexity. The administration staff are under the management of the company secretary.

EPR gets about 75% of its work from repeat business (both speculative work by developers and new contacts) and 25% of work from competitions and new contacts. The EPR principle is quality in all things — design, administration and in the achievement of buildings and interiors which meet client and occupiers needs in every aspect.

IT at EPR is organized centrally under the IT manager, Paul Slaney, who is responsible for all hardware and software. He is also responsible for the IT manual covering file drawings, use of CAD and the EPR way of doing things, e.g. fonts, title blocks, etc. The IT manager reports to the director with responsibilities for EPR resources. His view of IT is that it is a means to an end — 'if a practice does not use IT it just does not get jobs, often IT is client driven'.

IT applications for text manipulation, databases and specification writing

The EPR practice uses Microsoft Office on Windows NT platform; Word, Excel for door schedules, finishes schedules and charts; PowerPoint is occasionally used for presentations; similarly Access is also occasionally used for client lists and room schedules. The firm uses Specification Manager marketed by NBS Services and NBS on hard disks. Regular updates give the practice access to

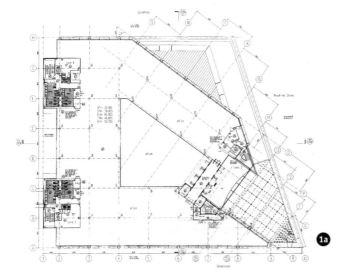

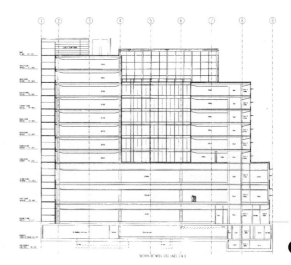

up-to-date information.

Accounts, although done separately on a separate UNIX system, are to be integrated on the Windows NT network. Staff however still fill timesheets manually before passing these to the accounts section for computerization.

Other software includes QuarkXpress, Photoshop and Illustrator which are used by a separate graphics department (one person) for marketing brochures and other promotional material.

IT applications for CAD

The firm has the complete version of MicroStation including 3D modelling and rendering capability. Of approximately 80 architects and technicians at EPR, 50 are computer literate compared to only 3 of the 13 directors. Some architects tend towards site supervision, others deal with design alone, but flexibility is expected of all staff. There are 75 computer users with 65 work-stations.

The EPR process covers

(*a*)   site analysis and investigations using Ordinance Survey Maps (which are now provided by OS as computer-based maps) and concepts development during the early stages of the design process

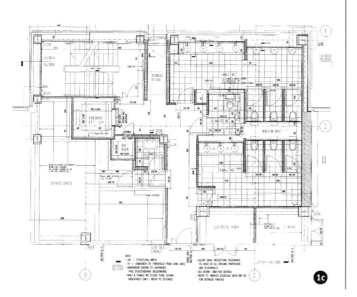

1c

(*b*)   2D drawings for layout development

(*c*)   3D modelling for massing and other functions.

The practice seeks to be proactive in the proper re-use of data properly and in the development of standards for drawings and all outputs (in particular developing a library of standard details). There is a heavy reliance on the use of physical rather than CAD models according to client preferences.

IT applications for telecommunications and networks

The EPR practice has a star ethernet network which is in the process of being changed to increase bandwidth. There is an ISDN line link with site offices.

A brief history notes that from 1982 EPR used to have a mini-based CAD system with CAD operators. The costs of running and upgrading the software and hardware and of employing CAD operators made the system obsolete and led to it being phased out. This was replaced five years ago with a PC network comprising mostly Pentium PCs (one 486 and Alphas, network plotters — colour and black and white). The policy is to use the fastest equipment for CAD workstations and the older equipment is then transferred for administration purposes. Internal E-mail is used for communicating with clients and other project team members.

IT training

The IT manager provides in-house training and arranges external training courses as needed. The IT department keeps up-to-date with technological developments through close relationships with vendors and specialist re-sellers as well as taking a close interest in the computer press both in print and via the Internet. External CAD training and support is provided.

Two issues of concern to the EPR IT manager are

(*a*)   that client driven IT demands are a problem

(*b*)   the speed of change of IT technological developments puts heavy demands on the practice to be dynamic and to adapt and change to keep up.

EPR addresses the first issue by using the IT department as a buffer between the client and EPR staff over the use of technology. The use of automated procedures can enable the EPR staff to see a familiar interface and work in the way they are accustomed while the client receives his data according to his specifications. Close communication is also important between client, architect and the EPR IT department as it enables the client's real needs to be clarified and the architects to be instructed on how those needs can be achieved within EPR's IT framework with as little additional cost as possible.

On the second point, EPR ensures that technology is specified for the sake of the practice's main business and not vice versa. By being cool headed about the claims of hardware and software vendors, especially over claims that their goods work as specified and that it is 'vital' to upgrade, the practice avoids technology which it believes is unnecessary or is used infrequently.

EPR tends to preserve lasting relationships with IT vendors. Such vendors take a long-term interest in what technology the company uses. It is then in their own interest not to sell EPR hardware which is unsuitable. With their inside information on industry development they are in a good position to recommend or even caution the practice against the purchase of certain technologies.

Selected projects

Headquarters of ABN AMRO Bank, 25 Bishopsgate, London

This project was undertaken with developers Spitalfields Development Group. Contributing to ABN AMRO's decision to locate the UK HQ in London was a design to match their business need, whilst at the same time providing them with the form of external expansion they felt appropriate to their position as a leading bank.

1. (a–c)2D plans, sections and elevations indicate the way CAD and IT are used in the design of a corporate headquarters
2. 3D illustration helps clients visualize the proposals

Eland House, London

This building, which has been designed to house the DOE, brings together the long-term needs of the building owner, Land Securities

plc and the Ministry. 2D plans, sections and elevations indicate the way CAD and IT are used in the design of another office building.

3. 3D illustrations indicate the use of CAD as a design tool. The façade is designed to maximize natural light and reduce solar gain thereby lowering cooling needs and energy consumption. This combines with innovative engineering including the use of a combined heat and power plant, to create a green urban building in keeping with DOE's national responsibilities

Fitzroy Robinson, Architects — 10

Background

Fitzroy Robinson was founded in the mid 1950s by Fitzroy Robinson and currently employs 100 staff (approximately 40% architects, 40% technicians and 20% support) under ten directors with offices in London (80% of all staff), Bristol, Cambridge, Glasgow, Budapest and Warsaw.

The firm's web page (*www.fitzroyrobinson.com*) notes the practice philosophy

- to achieve excellence in design and to provide the best professional advice to clients
- to further the art and science of architecture
- to provide a good environment and career for all people working in the Fitzroy Robinson offices.

The practice sees itself as distinguished by the fact that it creates contemporary buildings that serve both client and setting, meeting commercial and functional targets whilst contributing to local quality. The range of work carried out by the practice covers: commercial (offices), hotels and leisure, retail and master planning with clients including Barclay Group Property Services; BKD Bank; Banque Nationale de Paris; Bovis; British Aerospace; Harrods; English Partnerships; KPMG; London Underground; Norwich Union; Railtrack; Sun Life; Wates City of London Properties; Zeneca, etc.

Recent key projects are the Central Business Centre, Budapest, Barclaycard HQ, Northampton and BT Headquarters, Glasgow. Directors have overall responsibility for projects with project architects managing the projects as well as supervising architects and technicians according to the size and complexity of the project. A project CAD co-ordinator (one of the architects/technicians on the project team) is appointed for each project.

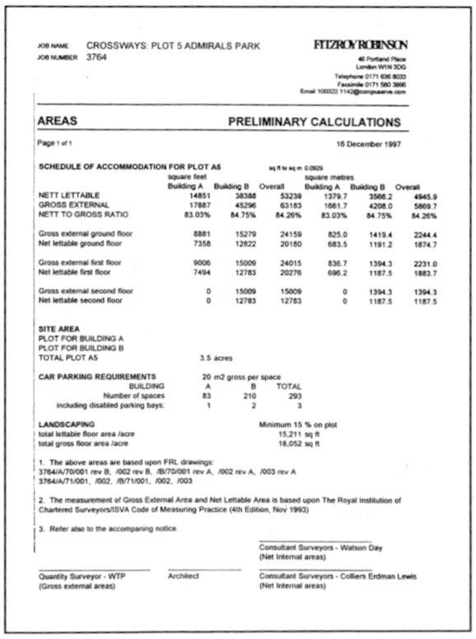

JOB NAME CROSSWAYS: PLOT 5 ADMIRALS PARK
JOB NUMBER 3764

FITZROY ROBINSON
46 Portland Place
London W1N 3DG
Telephone 0171 636 8033
Facsimile 0171 580 3996
Email 100322.1142@compuserve.com

AREAS — PRELIMINARY CALCULATIONS

Page 1 of 1 — 16 December 1997

SCHEDULE OF ACCOMMODATION FOR PLOT A5 — sq ft to sq m 0.0929

	square feet			square metres		
	Building A	Building B	Overall	Building A	Building B	Overall
NETT LETTABLE	14851	38388	53239	1379.7	3566.2	4945.9
GROSS EXTERNAL	17887	45296	63183	1661.7	4208.0	5869.7
NETT TO GROSS RATIO	83.03%	84.75%	84.26%	83.03%	84.75%	84.26%
Gross external ground floor	8881	15279	24159	825.0	1419.4	2244.4
Net lettable ground floor	7358	12822	20180	683.5	1191.2	1874.7
Gross external first floor	9006	15009	24015	836.7	1394.3	2231.0
Net lettable first floor	7494	12783	20276	696.2	1187.5	1883.7
Gross external second floor	0	15009	15009	0	1394.3	1394.3
Net lettable second floor	0	12783	12783	0	1187.5	1187.5

SITE AREA
PLOT FOR BUILDING A
PLOT FOR BUILDING B
TOTAL PLOT A5 — 3.5 acres

CAR PARKING REQUIREMENTS — 20 m2 gross per space

BUILDING	A	B	TOTAL
Number of spaces	83	210	293
including disabled parking bays:	1	2	3

LANDSCAPING — Minimum 15 % on plot
total lettable floor area /acre — 15,211 sq ft
total gross floor area /acre — 18,052 sq ft

1. The above areas are based upon FRL drawings:
3764/A/70/001 rev B, /002 rev B, /B/70/001 rev A, /002 rev A, /003 rev A
3764/A/71/001, /002, /B/71/001, /002, /003

2. The measurement of Gross External Area and Net Lettable Area is based upon The Royal Institution of Chartered Surveyors/ISVA Code of Measuring Practice (4th Edition, Nov 1993)

3. Refer also to the accompaning notice.

Consultant Surveyors - Watson Day (Net Internal areas)

Quantity Surveyor - WTP (Gross external areas) — Architect — Consultant Surveyors - Colliers Erdman Lewis (Net Internal areas)

Fig. 1. Fitzroy Robinson use Excel for schedules

Ross Gates (associate director), head of information technology, controls IT throughout the firm's five UK and two European offices with a strategy to reduce and exclude the use of paper and make more IT tools available at the desktop to improve efficiency of the design process. He is the author/auditor of the practice's CAD/project procedures in using Bentley Systems' MicroStation.

As chairman of the RIBA IT committee, Ross Gates is committed to promoting the use of IT to improve quality, efficiency and the range of services offered by members to their clients. Issues of concern include CAD procedures, benchmarking, live-rolling documents and the legal liability implications of electronic data transfer, for which a best practice policy and procedure for electronic data transfer of CAD vector data has been available since spring 1998.

Fitzroy Robinson is a member of the CICA Major Architects Group which includes BDP, EPR, Foster & Partners, Scott Brownrigg & Turner, Shepherd Robson and SOM.

IT applications for text manipulation, databases and specification writing

The practice has used Microsoft Office although the Smart II package remains in use. Word is used primarily for reports, letters, certificates, memos, minutes and specifications utilizing NBS templates, while Excel is used for schedules, such as for the accommodation areas for Crossways: Plot 5 Admirals Park development and software tracking audits (Fig. 1). PowerPoint is used for presentation but increasingly web site presentation is favoured because of the freedom it offers.

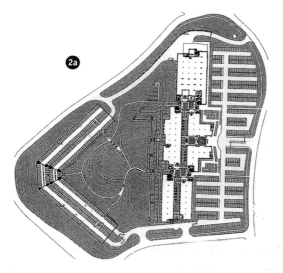

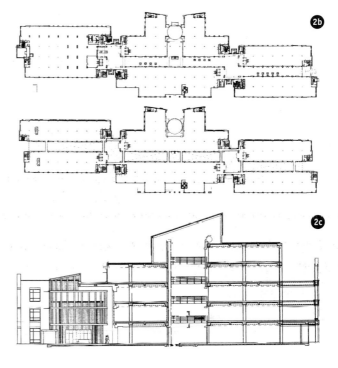

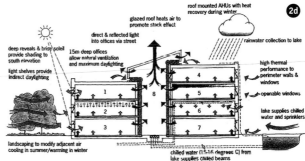

Microsoft Project is used for project planning and Schedule
Publisher for more sophisticated programming. Photoshop and
QuarkXpress are available for producing brochures within the firm's
graphics department. Accounts are processed employing an old
system on a VAX server but there are moves to replace the system
with fully integrated Windows-based software. Timesheets are still
completed manually although the practice is considering
integrating/interfacing with its electronic diaries through MS
Outlook to allow more electronic inputting.

IT applications for CAD

Since the early 1980s Fitzroy Robinson has had an in-house CAD
system known as DRUID which although still available is no longer
used for new projects. For these, the firm uses Bentley Systems'
MicroStation SE which now integrates raytrace and animation — the
functionality of Bentley's Masterpiece product. Fifty software
licenses on a concurrent licensing agreement are utilized by both
architects and technicians to produce 2D layouts and 3D work. 3D
visualization is generally outsourced to GMJ and Hayes Davidson
among others for broadcast and presentation quality.

IT applications for telecommunications and networks

The practice have a CAT5 network with 60 NT workstations linked to an
NT server. The firm's intranet, available via dial-up, contains central

resources and details such as
telephone and client lists.

IT training

All basic training is carried
out externally. Five IT
systems supervisors (under
the direction of the head of
IT) provide day-to-day
support to systems and the
IT help desk. More focused

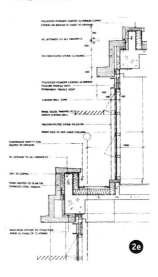

training is provided in twice weekly one-hour sessions available to all members of staff.

Selected projects

Barclaycard Headquarters, Northampton[1]
This building is a major low-energy HQ for 2000 staff.

2. (a–e)Plans and sections show formal elements consisting of long shallow blocks oriented east–west, shallow floor plates, atria or glazed streets which serve as environmental and social spaces, and mixed mode ventilation supported by air-conditioning in 'hot spots'. The plans indicate how the £42 million project uses IT to enhance the design process related to Fitzroy Robinson's architectural practice philosophy

BT Headquarters, Glasgow
3. In the BT Headquarters project a 3D model of the atrium space was produced during the early design stages and used as a tool

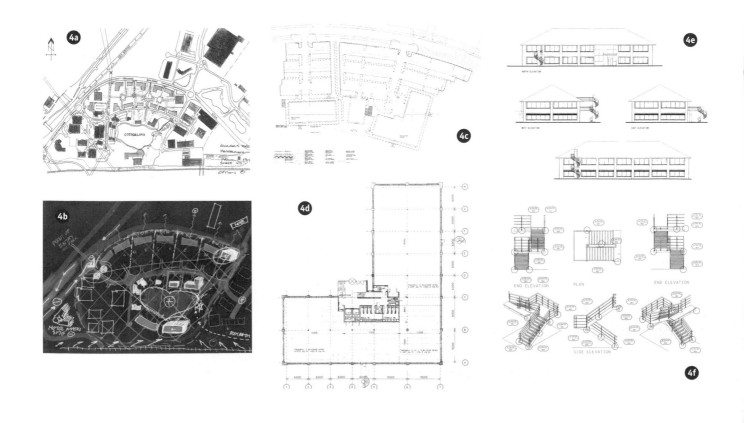

5a

5b

to analyse and track the behaviour of the sun (shading, screening roof, structural roof form, etc.)

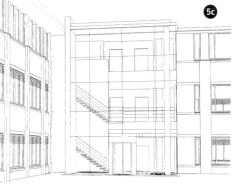
5c

5d

Office development at Admirals Park, Crossways

This development was undertaken for Whitecliff Properties.

4. (a–f) Plans, sections, elevations (and 2D layouts) show how the project benefited from the use of IT to complement traditional skills throughout the various stages of the design process from master-planning of the site to working drawings, including setting out and landscaping. Computers also facilitated detailed information on staircases

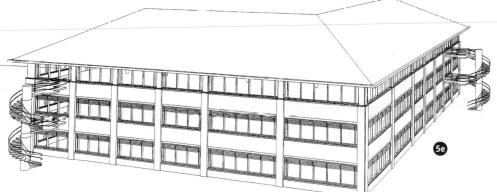
5e

5. (a–f) The 3D computer-generated images were used to examine options, and evaluate views and they provided the basis for hand-coloured perspectives

5f

Reference

1. Edwards B. Barclaycard headquarters: Green goes mainstream: Fitzroy Robinson. *Architecture Today*, 80, July, 1997, 20–36.

Gensler Architecture, Design and Planning Worldwide | 11

Background

Gensler has been listed in *World Architecture* as the largest architectural firm in the world (over 1400 employees worldwide).[1] Additional ratings by other trade publications acknowledge the firm as both the largest architectural firm and the largest interior design firm in the US. In an independent survey published by *Design Intelligence* (1997), a newsletter of the design industry, Gensler is recognized for consistent quality (of its design), professional and peer recognition, award-winning designs, most-respected design firm, best-managed design firm, loyal client base (85% return rate), staff retention, and financial health.[2]

Gensler was founded in 1965 and is now a broad-based organization with sixteen regional offices located throughout the world. At the heart of the organization is a progressive management, which is combined with teams of people who share a common philosophy and common goals in an organization which prides itself on 'flat hierarchy' (Fig. 1).

The firm is known for its specialization in architecture, interior design and planning services. It enjoys an established presence in the design of corporate or regional headquarters for blue chip companies as well as offices and fit-outs for professional services firms and high-technology companies, for building design and refurbishment, airports, retail stores and studios for the major entertainment firms. Its workload currently comprises 60% architecture, 31% interiors and 9% other consultancy projects. The senior governing body is the four-

person Board of Directors which is responsible for overall corporate policy and long-range planning (M. Arthur Gensler — San Francisco, Edward C. Friedrichs — Los Angeles, Margo Grant — New York, Antony Harbour — London Offices). The Management Committee, composed of thirteen members, sets regional operational policies and implements the Board's planning strategies. An Executive Council comprises taskforces and steering committees that address issues critical to the firm and the practice.

To ensure closer interaction with its clients, the London office opened in January 1989 at the invitation of Goldman Sachs and became a full regional office in 1993.[3, 4] It now has a staff of 120 professionals. Managing Director of its London practice is Tony Harbour, a British citizen and a founding member of the firm's Board of Directors and Management Committee. Today, its clients include Cazenove & Co., Linklaters & Paines, Credit Suisse, First Boston, Unisys, People Soft, BP, British Gas, Gap Inc., NatWest Markets, Stanhope Securities, Sun Microsystems, Canary Wharf Ltd, Thomson Travel Group, Nestlé UK, Development Securities, Liverpool City Airport, Royal Sun Alliance, McKinsey & Company, Warner Bro. and Walt Disney.

Strategically, the basic unit of organization is the project team — an assembly of professionals selected for their experience, capabilities and availability relative to a specific project. Central to the firm's success is that Partners actively participate in each project undertaken by the firm and although the firm may undertake any given number of projects at one time, each client receives the personal attention of the team of professionals dedicated to a project from its inception to its completion. As a firm, Gensler is concerned with building enduring relationships with its clients and, therefore, excellent service and attention to detail are its abiding trademarks. A measure of the firm's success lies in one telling statistic: 'more than 85% of its new business comes from repeat clients or referrals'.

IT policy and direction is set by the Gensler Board of Directors and Management Committee. Day-to-day responsibility lies with a team of IT professionals in each regional office whose efforts are led by a Partner currently located in the firm's corporate headquarters in San Francisco. IT managers in the

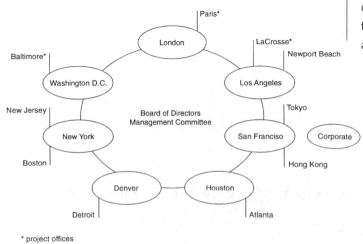

* project offices

Fig. 1. Office organization

regional offices meet at least once a year for a week to discuss policy and issues and they participate in regular conference calls for knowledge sharing and information exchange. Peter Claxton is the IT manager of the London office.

From the beginning, Gensler embraced IT as an inevitable and beneficial tool of the future. An early and strategic decision included a substantial investment in Intergraph CADD technology coinciding with a mandatory professional training program so that all professional staff, from the top down, from designers to project architects, would be trained in the use of the latest available CADD technology.

IT applications for text manipulation, databases and specification writing

Gensler's London office uses Microsoft Office: Word for specifications with reference to NBS for Windows, faxes, letters, memos; Excel for accounts reports, building studies, templates, drawing issues; PowerPoint for slides presentation; Access is partly used for marketing and tracking IT facilities management. First Class package is used for E-mail.

Microsoft Project is used for planning purposes and timesheets are produced with Time Keeper which is available in all offices linking to Harper and Schuman accounts software. Adobe Photoshop/QuarkXpress are used for working with images, rendering and marketing. FreeHand Macromedia is used to produce long banners.

Windows NT is available on two machines and it is considered that converting completely to Windows NT would expose the firm to security risks. There are four Windows NT servers in London.

IT applications for CAD

Gensler has approximately 540 CAD terminals firmwide and 440 PCs (124 machines in London). Ninety percent of office design projects involve the use of CAD. The technology infrastructure is designed to support the firm's vision of design and the performance of practice areas. Gensler

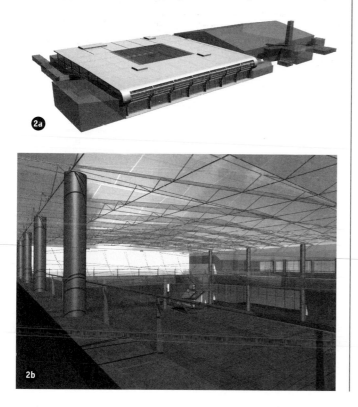

2a

2b

2c

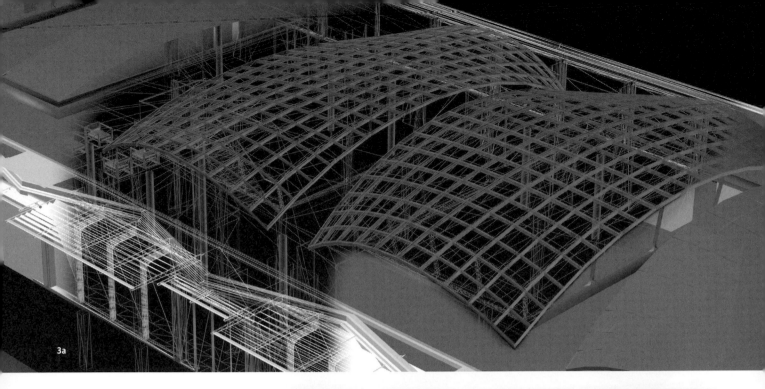
3a

uses CAD, 3D modelling and multi-media to enhance the design process as design tools, to inform clients and to do document work effectively. The creation of a predominantly electronic environment is seen as speeding up decision-making, strengthening communication and information sharing (to respond to emerging global needs), and reducing paperwork.

3b

In London, all machines are now leased and are by Dell, the preferred platform. MicroStation is the preferred CAD system although other offices such as San Francisco use AutoCAD. Other software packages available in London are 3D Studio Max for 3D rendering, Form Z and ModelView for modelling and 3D rendering. External agents are used for high resolution 3D visualization.

IT applications for telecommunications and networks

Gensler sees technology as a leveraging tool.

Technology has become an indispensable tool, the central nervous system of our firm. We now provide services of all types and scale to support national and international clients across our worldwide network. We improved our E-mail and voice mail systems, upgraded phone systems, located technology at each employee workstation, added support personnel and intensified learning programs. Through

on-line connections (e.g. BCQ on-demand) with many of our consultants, vendors, strategic allies and even competitors, we are forging new ways to ensure immediate customer response.

IT in the London office started with six Unix CAD and NFS networked stations using Intergraph on two Sun silicon graphics machines. There were four IBM PCs for administration. In 1993 a move to new offices in Roman House allowed development of the Novell PC network with two standalone CAD stations for a specific project over the six Unix CAD

3c

machines and network. By 1995 the office started the migration away from the Intergraph and Unix to PCs and a decision was made to buy PCs.

Although not having a web site, Gensler's intranet serves as a firmwide resource for Gensler employees, assisting them with project and marketing information.

A fount of data on teamwork, leadership, negotiation and presentation skills, project and time management, marketing materials, benefits including profit sharing and employee stockownership and media coverage.

Access to the Internet is by Microsoft Internet Explorer. The same level of service — same phone, WAN, E-mail and voice mail for communication at every workstation throughout every office is seen as crucial to promote a service-oriented culture. Gensler has four main servers (East, West, Europe and Far East) to facilitate communication.

Individual groups across all the offices come together — through conference calls, intranet postings, task force meetings, newsletters — to generate the most cutting-edge information and approaches available. Interactive data conferencing will allow design charrettes via computer, across continents, and among Gensler offices, clients and consultants.

IT training

At the heart of the Gensler training organization is the 'Gensler University' for senior staff, who can enrol for two year training programmes in management, firm values and skill development. The university is also charged with developing training programmes for individual offices and individual studios to improve their knowledge-building capacity.

The firmwide training programme (with an IT budget) allows in-house CAD training by co-ordinators and external non-CAD training. There is also a CAD test and compliance with standards defined throughout the firm. A CAD manual is available in the US offices with the London CAD committee directly responsible for IT standards in the UK.

As space and streamlined furniture standards become the norm and integrated CAFM systems — that not only track furniture inventories and life-cycle costs of building systems but link this information to human resources and accounting departments — begin to emerge, it is clear that the architect and designer of the future will be part designer, part strategist, part social engineer, part manager.

Selected projects

These are centred on a Call Centre for a large industrial client, a campus-based and a city-based corporate headquarters. All the images illustrated in this publication were modelled in-house using MicroStation and rendered in MicroStation or ModelView in various combinations. They then went through post-production using Adobe Photoshop 4.

Call Centre for a large industrial client

2. (a–c) The illustrations show the way in which Gensler was able to demonstrate how the interior could be cleverly utilized to the client's benefit by judicious space planning. The 3D images from the inception and design development stages were very useful during presentations to the client

4d

Campus-based corporate headquarters

This project for a financial services firm involves restoration of a Grade II listed neo-classical mansion, and the design of a new building to accommodate trading floors and support areas. It is located on an eight acre site. Once restored the manor house will be used as an entertainment facility with hotel-quality suites, the servant's quarters will house office spaces and a commercial-type kitchen. The design also comprises an additional 2580 m² single-storey trading facility and a gym with an indoor pool at the back.

3. (a–c) Plans show how Gensler uses IT to offer some options to meet client expectations. The 3D images from the design development and preplanning stages were used to explore options and to express the design intentions to the client. These show external perspective views of the development

City-based corporate headquarters

The 17 760 m² building for a UK-based international financial institution involved producing a strategic brief covering headcount projections to the year 2008 and accommodation comprising executive offices, client conference rooms, dining facilities, trading floor, general office leisure and recreational facilities. Whereas the lower ground floor plan includes a gymnasium, reprographics and sandwich bar, the ground floor plan accommodates the main reception and additional office spaces. The top floor (8th) contains the executive and client areas and includes four private dining rooms, two video conferencing rooms, one boardroom for

up to 26 people, one multi-function room with a demountable wall to hold between 16 and 120 people and various additional meeting rooms and a boardroom for up to 42 people. The building was a speculative development with a distinctive design, achieved by way of branding and the use of graphics to reflect the clients' preferred finishes. The interior design of the general office addresses the main issues of maintenance to minimize wear and tear of furniture and floors, and to accommodate the high turnover of staff by implementing the 'universal plan' which utilizes simple, generic desk furnishings and multi-use spaces.

4. (a–d) The plans emphasize networking and communications facilities. They also depict the main reception area in the final design development stages. On seeing the images of the main space the client requested a more detailed study of the finishes exploring different options. The 3D images facilitated the design and planning

References

1. Gensler A. Gensler Architecture, design & planning worldwide. *World Architecture,* No. 39, 1995, 40–81.

2. Gensler Annual Report (illustrated by James Steinberg). 1997.

3. Russel B. Gensler goes international: Antony Harbour heads the new regional office in London. *Interiors,* February, 1994, 35–53.

4. Rus M., Davidsen J., Geram M. and Bussel A. Gensler: anatomy of a giant — the business of a design. *Interior Design (special edition)*, November, 1997, S5–S139.

HOK International Ltd, Architects	12

Background

Hellmuth, Obata and Kassabum, Inc. is a full-service design organization with more than 40 years' experience. It is a diversified practice including comprehensive architectural, engineering and interiors services as well as services in a wide range of related disciplines from facilities consulting to development of master plans. The staff of over 2000 professionals is experienced in a wide variety of assignments for major corporations, developers, state and local agencies, sports facilities, hospitals, colleges and universities, the US government and governments in Canada, the Caribbean, Central and South America, the Middle East and Asia. HOK has offices in Atlanta; Berlin; Chicago; Costa Mesa, CA; Dallas; Greenville, SC; Hong Kong; Houston; Kansas City; London; Los Angeles; Mexico City; Moscow; New York; Orlando; San Francisco; Seattle; Shanghai; St. Louis; Tampa; Tokyo; Toronto; Warsaw and Washington DC.

The most important factor in the success of HOK has been its ability to manage the total planning, design and construction process for projects of any size, type or scope, located anywhere in the world, along with its record of delivering projects on time and within budget. In addition HOK is committed to design excellence and quality of service.

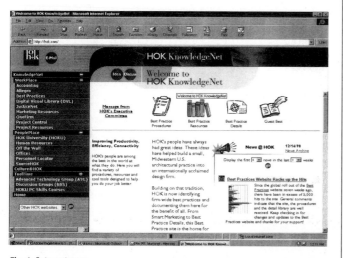

Fig. 1. Intranet page

HOK opened its London office on the back of a major project — the world headquarters of BP in Finsbury Circus. The firm merged with Cecil Denny Highton (a practice with a reputation for preserving and recycling buildings) in November 1996 and operates as HOK International Ltd. The managing director in London is Ralph Courtenay. The firm serves clients from the Far East and South Africa, dealing with its Birmingham Centenary Plaza Scheme alongside work from Portugal and Russia. HOK has a number of high-tech clients including Cisco, Nortel, Microsoft, Sun Microsystems and BAA plc.

The range of work covers office buildings, mixed-use and retail facilities, hotels and resorts (hospitality), and interiors. HOK's conservation group has been involved in historic landmark projects such as Houses of Parliament, Treasury, the Foreign and Commonwealth Office, and the Natural History Museum. HOK strives to fully understand organizational business goals and strategies in order to design quality, efficient spaces that enable people to do their jobs well.

The firm aims for a global community — one firm. Effective IT and communications are crucial to achieving this. In practice, this is achieved by linking offices to the St. Louis Corporate headquarters via a wide area network (WAN). The immediate task becomes one of

(a) integration — 'knitting HOK's diverse constituent practices'
(b) standardization — developing standard approaches to CAD, details, office procedures and conditions of employment
(c) simplification into a distinct HOK approach and support.

Often this is done through a series of guidelines and using focus groups.

In London, the use of computers started around 1980 with the purchase of an Apple II for design calculations and office administration. CAD systems followed with a major investment in the purchase of a GDS system (then £180 000 for a three workstation system). A further two generations of minicomputers followed before the system was phased out in preference of PC-based systems. HOK has now achieved its aim of a PC on every desk and only a few

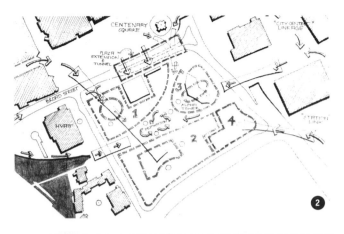

drawing boards remain. The London offices now have over 60 PC workstations. The firm regularly purchases high performance CAD machines and operates a 'trickle down' policy whereby these are reused for less demanding marketing and administrative functions. The current specification is 300 MHz Pentium processors with 64 or 128 MB RAM.

Architect Terry Nichols, HOK's IT Manager in the UK, explains that standardization and simplification are the two main issues.

IT is both a servant and master: that is, servant in facilitating global communication or fast delivery of project information; and master in driving change by continually extending the palette of options.

Individual projects are run by project managers and project architects who are in charge of a team of architects and technicians with project directors having overall responsibility.

Rather than a CAD manager for the whole office each project is resourced with a CAD co-ordinator to plan, standardize and develop specific needs for the particular project. Increasingly there may be a new role to co-ordinate the individual project CAD managers.

IT applications for text manipulation, databases and specification writing

HOK use Microsoft Office: Word for word processing for letters, memos, minutes and E-mail; Excel for financial and design calculations. PowerPoint is increasingly being used for presentations. A number of management databases have been developed using MS Access. Outlook is the preferred client for E-mail. Pagemaker and QuarkXpress are used for marketing information, brochures and project bids. Specification is done using NBS which is on the server.

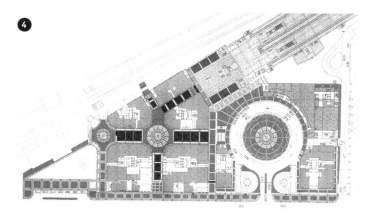

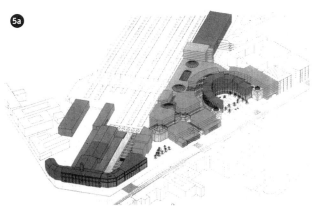

Accounts are done using a US package called Advantage. This is linked to a resource manager tool (Allegro) for resource planning. Power Project by Asta and Microsoft Project are used for project planning.

IT applications for CAD

AutoCad and 3 D Studio are the main CAD applications although HOK is continuing to develop 'DrawVision', an in-house Windows-based CAD system for its particular strengths in preliminary design and 3D presentation work. The practice also uses other CAD software such as MicroStation in response to specific project demands. Terry Nichols notes:

Within the last five years, CAD work has changed from being mainly 2D layouts and now, increasingly, 3D drawings are the norm. We are currently experimenting

with VRML (Virtual Reality Mark-up Language) technology. A number of projects have used links from drawings to database records for schedules of accommodation, cable management records, furniture inventories and other facilities management purposes. There is the building of a library of best practice standard details as a database front end of the CAD system. A challenge is to standardize the way staff survey and record historic and other buildings so that information can be held centrally in a uniform way, for easy access.

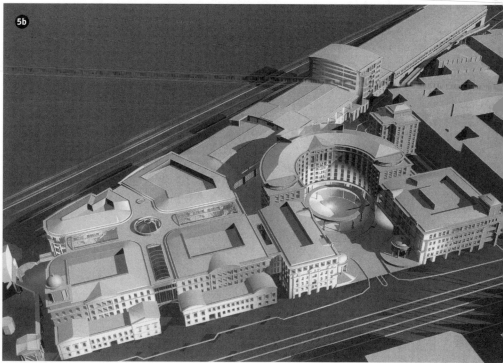

IT applications for telecommunications and networks

The practice runs a Windows NT network with Exchange Server for E-mail. The three locations in London have a total of six servers.

HOK have been developing the opportunities of the Internet, intranet and extranets.[1] The key is interchangeability. HOK's Internet homepage is located at *www.hok.com*. This includes projects, its design approach and career opportunities. The menu of other sites includes Europe, Asia and sport expertise. The HOK intranet (Fig. 1) is used to enhance the sense of the one firm with resources for marketing, standard details, publication of the work of inter-office focus groups and routes to local home pages of some individual offices. HOK also uses the Internet for the transfer of drawings. The drawings are loaded over a lease-line connection to the corporate FTP (file transfer protocol) site. Recipients are sent an E-mail to say the information is available for download via their own Internet connection. Another issue concerns increasing the bandwidth from 64 K at the moment as the practice wants the capacity to deliver projects anywhere in the world.

IT training

Staff IT training needs are regularly reviewed. HOK has a quality manager responsible for monitoring training needs and organizing courses.

Selected projects

Centenary Square, Birmingham

Hampton Trust and Carlton Communications, joint owners of the 4 ha city-centre site, have come together to create a significant leisure and entertainment development. Anchored by a new landmark tower, providing space for Holiday Inn, the scheme will regenerate a strategic city site, completing the pedestrian linkage from Brindley and Centenary Square to the city centre.

2. Plans, sections, elevations and working details are developed from freehand sketches and IT tools allow HOK designs to communicate with their clients
3. (a–b) 3D illustrations enhanced using CAD (courtesy of Hayes Davidson)

Moscow Terminal, St Petersburg, Russia

This project is the most significant development to take place there in recent years, representing a major milestone for the city in terms of regeneration and urban renewal. The mixed-use development of 120 996 m^2 over 4 ha consists of a rail terminal for the new high speed link between St Petersburg and Moscow, a new corporate headquarters, a 350 room hotel with business and conference facilities, six office buildings, serviced apartments, arcade shopping and parking for up to 800 vehicles. HOK's design, incorporating the existing Moscow Railway Station, built around 1843–1851, and the adjoining buildings, provides a strong urban statement while balancing physical constraints, financial realities and operational objectives.

4. Plans, sections, elevations and working details show how IT allows HOK designs to examine options that are cost-effective, market responsive and flexible to meet evolving tenant and user demands. IT is allowing work to be shared between clients and consultants in Russia, London, Basingstoke and US offices
5. (a–c) 3D CAD illustrations enhance visualization of HOK's design based in neo-classical inspiration

Designing Corporate Interiors

For example BP offices; Instinet UK Ltd, London and Glaxo Wellcome Research & Development, Stevenage. In such projects IT allows HOK designers to communicate with their clients and to keep up to expectations of tenants.

Terminal 5 Heathrow Project, London

For this project the consulting group, principally made up of architects and often representing groups from other disciplines — sociology, psychology, etc., played a crucial role in constructing a dialogue between British Airways and the British Airports Authority — a process of 'defining a client's goals and needs before concept design commences, in contrast to a less structured approach of producing sketch schemes to establish a brief.

Reference

1. Evans B. Unravelling the web at HOK. *Architects Journal*, 23 October, 1997, 51–2.

Ian Ritchie, Architects | 13

Background

Ian Ritchie Architects was formed in 1981 in London by Ian Ritchie when he was director of Rice Francis Ritchie (RFR Design Engineering) in Paris with Peter Rice (engineer) and Martin Francis (naval architect/industrial designer). Prior to this Ian Ritchie was project architect for Foster Associates and also acted as a design consultant for Michael Hopkins Architects (SSSALU), Ove Arup and Partners and Peter Rice (Shelterspan Fabric Structures).

Ian Ritchie's book[1] describes the way in which the practice approaches architectural design and the execution of each project, tailored by the original preconcept: a holistic expression of the project's essence — informing and providing the framework for developing the architectural design. As Ian Ritchie notes the book is 'not about my architecture, but the way we, in our studio, think and proceed collaboratively in realizing it.' The book shows the practice aims.

The importance of scale shift, the perception of the overall form and increasing the viewer's awareness that a geometric hierarchy is the basic compositional element of our work began to appear clearly through our work with Peter Rice and Martin Francis on the La Villette facades and has been continued through the B8 Building at Stockley Park, Reina Sofia Museum of Modern Art, (Madrid), and Magdalen College.

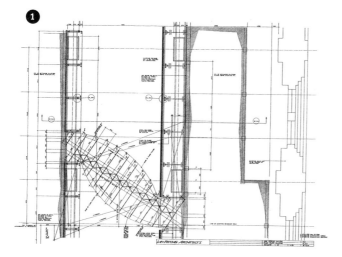

The practice is about Ian Ritchie's preoccupations, for example that with light. The essential material of architecture is manifested in the glass facades of the La Villette Museum of Science & Technology, Paris. In architecture, glass has for a thousand years been the medium through which light has entered buildings. The possibilities of glass were researched in order to achieve transparency, the invisible. Here, a small-scale, flush fixing component was developed which was able to take several times the previous proven strain.

The firm consists of fifteen architects and three supporting administrators. There are no technicians. Four senior architects act as project managers. The size and organization of the design office is important and the number five allows a group of individuals to work well on a design. The practice is made up of five such groups directed by associates handling the larger projects. About two dozen staff seems quite common in practices led by a single principal. Ian Ritchie is concerned with turning pre-conceptual ideas into the actual building. These are the primary generators.[2] The firm seeks

to use technology for the sake of creating an image rather than for its own sake. It values tools, traditional drawings, the computer models and the physical models for what each brings to the design process.

IT management is shared between administrative staff and an architect with an MA in Computing, while IT requirements are widely discussed within the whole team. Once a week, Ian Martin Associates Consultancy helps develop the IT strategy and supports the day-to-day operation. The firm is keen that IT should take the 'cumbersomeness' out of activities especially where there is repetition, allowing structuring and systematic procedures, exploration of ideas and design solutions.

In many ways the perception is that IT should help preserve the status of a small practice (20 maximum) comprised of interested

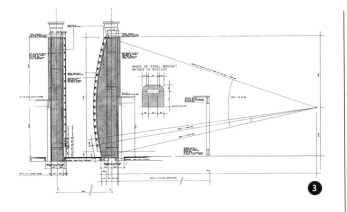

There is no IT budget although the requirements are project-driven depending on workstation needs to serve both conventional and innovative projects as well as the needs to upgrade and replace hardware and software. Nonetheless, priority is given to keeping IT equipment and software at the highest performance level.

IT applications for text manipulation, spreadsheets and databases

Ian Ritchie Architects use Microsoft Word for word processing (an upgrade from 'Write Now' software). They also use this to type up specifications with reference to hard copies of the NBS documents. Excel is used for standard forms and templates, for drawing issues and for accounting for which they find Excel to be easy to use and versatile. File Maker Pro is used for making lists of contacts while QuarkXpress/Photoshop are used for brochures and presentation material.

IT applications for CAD

The firm has the full version of MicroStation including 3D elements which replaced the Gable CAD System (from the University of Sheffield). The hardware consists of Apple Macintosh machines — fifteen

individuals who are happy to work together and who share belief in a similar philosophy of life. This includes preserving the way the office functions, 'that territories do not have boundaries, they are simply different landscapes which require different skills to negotiate well, but also through which one's collaborators can be supported and supportive.' This means IT aiding the achievement of practice aims to reap desired benefits of collaboration, whether in relation to the user or industry; or a better, more valid and flexible architecture achieved by access to scientific and technical knowledge, breaking down barriers between disciplines, and developing greater awareness of social and ecological issues.

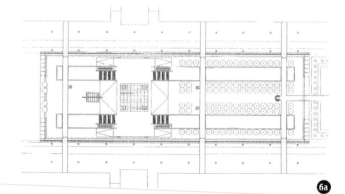

the office to lead a project within an atmosphere of mutual support, and collectively to explore ideas, materials and techniques in the knowledge that research, application and evaluation will be realized relatively quickly. Both as suppliers of hardware/software and as service providers, Ian Martin Associates Consultancy are available to give any support and training as necessary.

PowerMacs (one 7100, one 7200, four 8500, five 9500), one Quadro and one LC; A4/A3 laser printers, A0 colour plotter; A0 black & white Hewlett Packard plotter; one server. One of the machines is used for administration purposes.

The practice uses folders to structure digital information, for example surveys, layout and 3D details. Ideas and concepts are translated into CAD drawings with exploratory models for setting up perspectives.

IT applications for telecommunications and networks

Ian Ritchie Architects have an ethernet network and an ISDN link. On security aspects, back-ups are carried out on a daily basis and it is usual to double-up, that is issue electronic data followed by paper copies. One fax machine is integrated within the network and is used to send and receive information; while another fax is separate from the network and is operated manually.

IT training

Ian Ritchie Architects have a policy of in-house training through experienced users on a day-to-day basis in order to avoid training sessions in which 'things are thrown at you'. This is in a framework which uses mechanisms of training and learning on small projects. Also this provides the opportunity for the less practised architects in

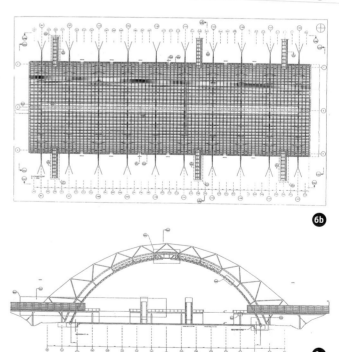

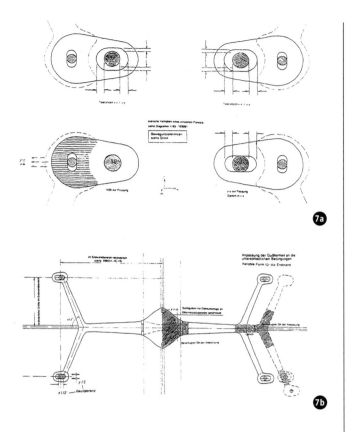

(7a)

(7b)

integrated two-dimensional drafting and three-dimensional modelling... allows the user to modify building, ground and road models together interactively. This meant at the Natural History Museum that to get the panels on a human scale the architects were able to try out many different sizes before making the final choice.[3]

The system also helped in deciding the geometrical configuration of the bridges. Should they be straight, curved or diagonal, and how should handrails be supported? The manipulation facilities allowed architects to draw 20 different possibilities quickly. 'Because the bridges have curves in both elevation and plan they would have been very difficult to draw by conventional means.'

1. In the plan of the ecology gallery CAD gave the opportunity to try alternative design ideas, to explore ideas (e.g. the complexity) in an attempt to answer the geometry brief for an exhibition enclosure that would reinforce the message of material on show
2. Cut away plan showing the design
3. Elevation hole through straight wall and other details
4. (a–b) 3D computer-generated illustrations of the bridges

Leipzig Messe Glass Hall 1992

The Leipzig Messe Glass Hall is more than 240 m long and 80 m wide with 1140 tonnes of glass suspended from a single-layer steel barrel vault. The project for the Leipzig Fair was designed by Von Gerkan

Key issues, for the practice, concern ensuring that each architect has a CAD workstation and that IT tools are up-to-date. This necessarily requires upgrading of hardware and software continuously. Also this must facilitate IT applications from the extensive simple use in 2D work to limited complex 3D modelling and rendering ('the seductive images produced by the special tools and skills'). IT tools must aid the process of meeting the client's requirements.

Selected projects

The Ecology Gallery at the Natural History Museum, London, 1990

The final design of the two-storey glass structure benefited greatly from the ability of GABLE's 4D series to create alternative designs quickly. The Apollo-based GABLE CAD Series 4D which was one of the first architectural systems to offer fully

(7c)

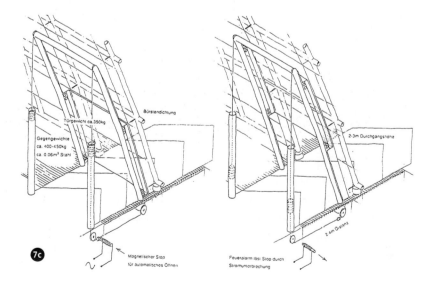

Marg und Partner in collaboration with Ian Ritchie Architects. Transparency was a key objective. Use of IT tools is enhanced by characteristics such as the shell's structural grid of 3.125 m and glass-panel size deriving from the 6.25 m Messe architecture module, size of the structure, repetition of elements, etc.

An analysis of symmetrical and asymmetrical loading, wind-tunnel tests, a study of snow loads and the relationship between wind and snow loads, and a full non-linear computer modelling of structural behaviour gave us predictions of deformations under load.[4]

5. Structural legibility and expression of the overall form are shown in the hierarchy of principal elements produced at the tender stage. The structural layers of the vault are clearly separated: the ten primary arches, the tubular grid shell, the thin cast-steel fingers holding the glass, and finally the glass layer itself, hung inside the structure in a single sweeping uninterrupted span

6. The ground floor plan (a), roof plan (b) and section (c) relate to the early design sketches of a barrel-vault grid shell outside hanging a continuous glass envelope below it

7. Drawings were extremely detailed (a) and included full size details and indicated the architectural and detail design intentions. There was also a focus on the detailed written specification. Details of 'fingertip' connections of the arms where a slot and radial movement with serrated interlocking edges were provided (b) as well as anchoring details (c)

8. Computer simulation of the structure indicates a means to evaluate design objectives

References

1. Ritchie I. *(Well) connected architecture*. Academy Edns, Ernst & Sohn, 1994.

2. Lawson B. *Design in Mind*. Butterworth Architecture, 1995, 83–91.

3. Ritchie I. Natural selection: how CAD helped Ian Ritchie design the new ecology gallery at the Natural History Museum. *Architecture Today*, 14, 51–2.

4. Ritchie I. *The biggest glass palace in the world*. Ellipsis London Ltd, 1997, 39.

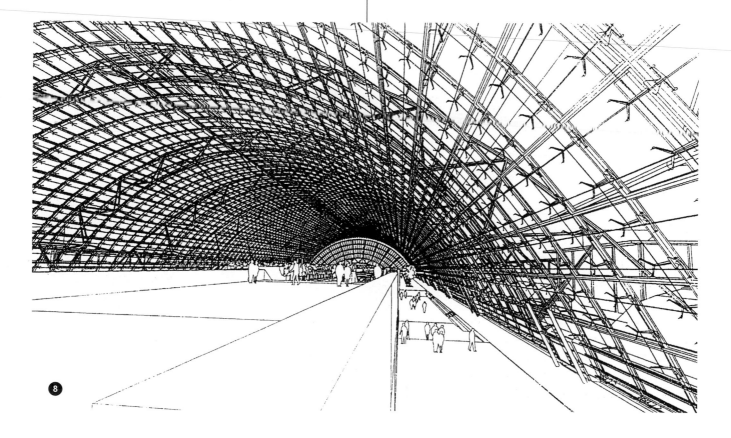

8

Kohn Pedersen Fox, Architects | 14

Background

Kohn Pedersen Fox (KPF) has two main worldwide offices, New York, and London, with Berlin and Tokyo as satellite offices. KPF was founded in 1976 as a US practice by Eugene Kohn, William Pedersen and Sheldon Fox

with the intention of creating architecture that would make positive and sensitive contributions to the environments of cities throughout the world, i.e. seeking

- to make buildings active and engage participants in the urban scene

- to put their stamp on the fabric of urban America and to create works of architecture that redefined and benefited their surroundings.

The London office[i] headed by Lee Polisano, senior partner, opened in 1989 with a staff of 43 architects and planners despite the recession, because of London's reputation as a financial centre and because KPF valued London as one of the three 'truly global cities' alongside New York and Los Angeles. For David Leventhal, Lee Polisano and his partners 'London has not only given us access to Europe, it has given us access to a lot of the rest of the world'.

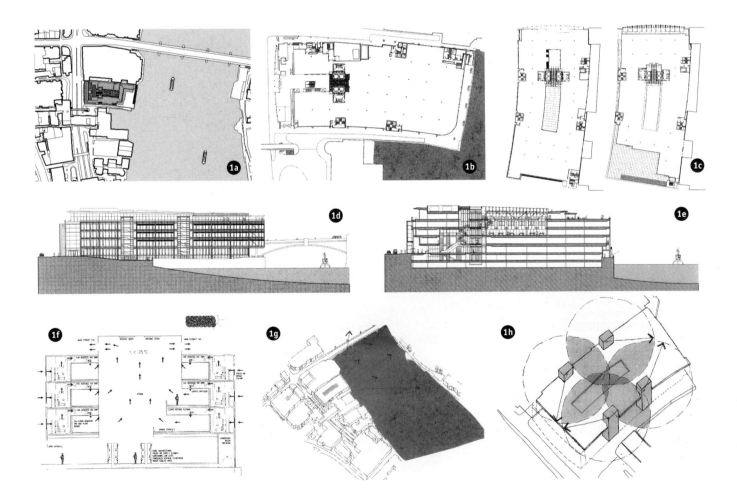

The nature of work undertaken, including masterplanning, comes mainly from the financial and the office sectors. The London office is also doing significant work outside the UK including Germany, the Netherlands and Cyprus. For any project in the office there is a project manager alongside a design manager both of whom are responsible to a senior manager who oversees a team of about 4/5–8/10 designers who are computer literate. Other than a wide variety of different computer tools the team typically use up to 200 clay models for a project during the design process.

IT applications for text manipulation, databases and specification writing

KPF use Microsoft Word for Windows for text manipulation, letter writing and report writing. They also use QuarkXpress/Photoshop for desktop publishing; the accounting system SAGE; the 'Project Command' for timesheets; spreadsheet 'No Teamate', and use Microsoft Word for specification writing.

IT applications for CAD

Interest in CAD dates back to 1983/4 when the New York office purchased the Intergraph System. Since 1985 KPF has stayed with the MicroStation industrial standard on PCs using Windows and

Windows NT. The New York office uses PCs by Dell while London assembles its own machines.

Lars Hesselgren (KPF) and formerly of YRM states that

As extensive users of MicroStation which we use for all aspects of our work — all working drawings, 3D modelling, animations, presentation plotting are done with this tool — we feel we are among the premier CAD users anywhere.

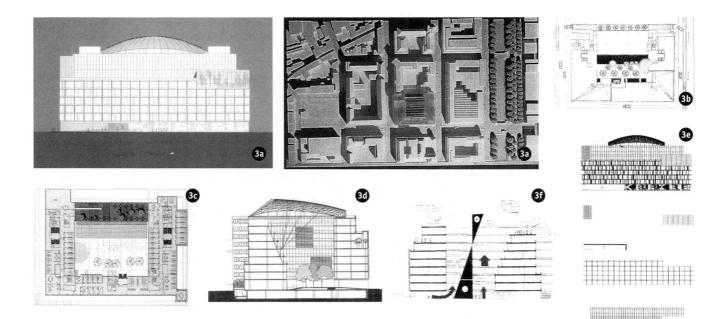

KPF use the specialized rendering software, Bentley Masterpiece, for presentation drawings. Lars Hesselgren is also Chairman of the MicroStation Community – the Bentley user group.

IT applications for telecommunications and networks

KPF has about 60 Pentium machines/workstations (40 of which are for CAD and 20 for administration); one plot server, one file server for storing all the data available through the network, and an e-mail server. Lars Hesselgren (KPF) notes

Our system is Win95/98 and WinNT and as you see we use MS Exchange Server for e-mail and our web site is under construction with development of an office intranet.

IT training

KPF have a policy of employing experienced CAD users. Staff can request in-house training, for example refresher courses on basic 2D drafting, advanced image manipulation, basic 3D modelling, or advanced image processing from Lars Hesselgren who also arranges and offers CAD half-day courses especially to new

employees. Three times a week IT training sessions focusing on specific topics are also held using external trainers.

Issues of concern cover the legacy data (the need to move data into current format using current media); the need to use tools which allow more utilization of 3D capabilities, and inter-operability within the project team allowing seamless exchange of information. The Consortium of Architects, of which KPF are members offers hope as a virtual centre which may well address some of the management issues to do with the integrated model proposals.

Selected projects

Thames Court Development, London[2, 3]

This project has a standard British financial institution's brief: deep plan trading floors and heavy servicing in a mixed use complex of 30 000 m^2. The use of the computer aids design development, environmental and facade analysis.

1. 2D CAD drawings of the site plan (a) ground level plan (b) level two plan and roof plan (c) show the speculative five-

storey office building occupying an entire city block. The building has a steel frame chosen to minimize works to existing foundations structures, with the whole centre of the building suspended from two trusses at roof level creating a large column-free area to the dealer floor at first floor level. The Queenhithe elevation (d) is shown responding to the pedestrian route and long section facing west (e). The air distribution strategy is shown in (f). The atrium roof shaded by motorized fabric paddles acts as an exhaust air plenum for the offices. Environmental analysis (g) is shown as determining the form. Although not a

completely naturally-serviced office building, the innovative idea is to recognize that during two of the year's four seasons, extra heating and cooling are almost certainly unnecessary. Locating services, safety routes and workshop are shown in (h).

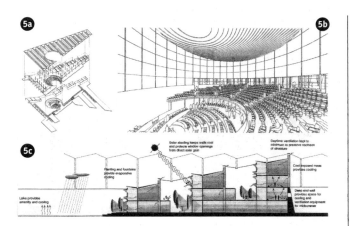

2. (a–e) 3D computer models built-up on MicroStation and computer images by external specialist firms such as Hayes Davidson complement physical models

KBB New Headquarters Building

The 30 600 m² and 26 100 m² rentable area in the heart of Dusseldorf's financial district has banks of offices arranged around a central atrium with ground floor retailing and a rooftop café. The design tries to incorporate some efficiency advantages of a deep plan with German norms — regulations set standards for the workplace enforced by white collar unions.

3. (a–f) 2D CAD and physical layouts of the site plan (a) ground floor plan showing area given over to facilities with public access (b) upper and lower floor plans (c) section (d) are used as a means to improve the office block. The courtyard acts as giant plenum — a thermal buffer between the exterior and the office environment. Facade treatment analysis (e) shows patterning of the facade generated by the fall of sunlight across the building with darker areas where extra shading might be needed. With the innovative servicing strategy, the earth is used to cool air which is then fed into the atrium. The courtyard acts as a giant plenum (image by HL-Tecknik) — (f) section showing servicing strategy

4. (a–h) 3D physical and computer modelling shows the optimum roof design, achieved by engineers RFR with curved glass over the atrium and movable sunshading components

The House of Representatives, Nicosia, Cyprus[4]

This is a new four-storey building of 20 000 m². The basement car park for 160 cars is notable for using an intriguing material —

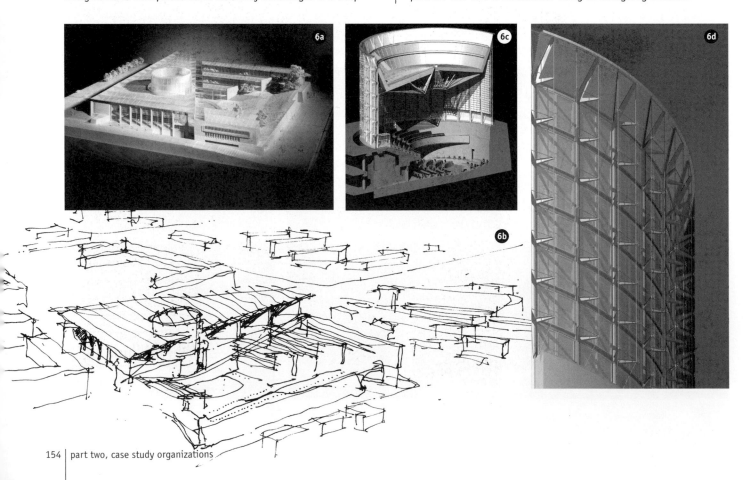

alabaster. A primary role of the building, won through a competition, is to encourage citizens to participate in the country's democratic process.

5. (a) Ground floor/upper floor and roof plan shown layered early section; (b) the circular debating chamber (c) cut away section showing the use of alabaster as a feature for the drum. The chamber rises above the foyer roof as a dramatic landmark on the horizon. Section shows the steel blinds casting shadows on the back of the alabaster.

6. 3D illustrations (CAD and physical models): (a) aerial view of the parliamentary garden. Offices are organized around small-scale garden courts. The gardens connect with a parliamentary park which offers members a place of quiet retreat; (b) freehand sketches of the building layout; (c) section; (d–h) internal CAD based views

References

1. *KPF International World Architecture Profile magazine*. No. 34, The International Academy of Architecture. 1995, 26–65.

2. *Ibid.*, 54.

3. Candell I. Deutsch marks the spot. *Property Week*. 4 April, 1997, 32–33.

4. *KPF International World Architecture Profile magazine. op. cit.*, 62–63.

Opel Kreisel Project

This project is a multi-building complex of 25 000 m², including provision for offices, financial services' trading floors, a cafeteria and below ground parking in Frankfurt.

7. 3D modelling computer images of the building

Leslie Jones, Architects | 15

Background

The practice was formed in 1945 by Leslie Jones and now has 80 staff under eight partners in two offices (London and Manchester) with Andrew P. Ogg, as chairman, having overall management responsibility, and one partner being responsible for the management in each office. Bristol and Poole branch offices have since closed.

Leslie Jones' portfolio of work is retail and leisure dominated — with clients such as Land Securities, Rank, ANC, owners of multi-complex cinemas — but includes business facilities, historic buildings, new build or refurbishment. The firm's expertise covers masterplanning, design, interior design, space planning and fit-outs, facility audits and evaluations. The firm's experience of retail design dates back more than 25 years from the design of its first UK covered shopping centre in Poole in the 1960s. Since then major shopping centres include the 28 800 m^2 The Galleries, Bristol for Norwich Union, Kingston Centre, Milton Keynes and Central Birmingham Retail Centre.

The firm is dedicated to achieving high quality standards in all facets of its service and is committed to the realization of projects within agreed cost, time and quality constraints. Each project is under the direct control and leadership of a partner. Larger and more complex projects may involve input from two or more partners, evolving and monitoring design and controlling technical production and construction. A senior architect acts as design team leader with responsibility for design team management, ensuring that policy and co-ordination between team members is properly implemented and sound project planning undertaken.

The practice has an environmental policy and has quality control procedures in line with BS 5750 for all aspects of the design and construction process. Office standards setting out desired approaches are communicated through IT and CAD manuals. The practice is committed to IT and believes 'once a practice takes the IT route — there is no alternative'. IT policy with the two offices aims for PC and Macintosh compatibility. Martin Perry, an architect with AutoCAD and MicroStation experience, has been with the practice for two years overseeing the significant IT development and (software and hardware) investment (about £250 000).

IT applications for text manipulation, databases and specification writing

Leslie Jones uses Microsoft Office: Word for word processing, letters, minutes, memos, and for brochures since the closure of their graphics section; Excel for forms, schedules, e.g. meetings, drawings

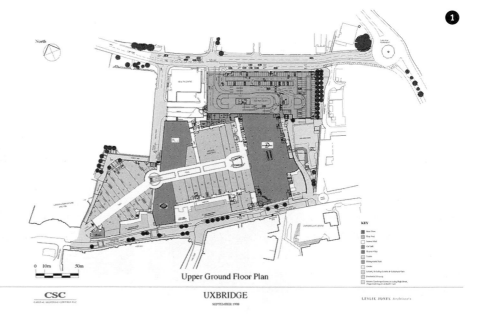

North

Upper Ground Floor Plan

0   10m   50m

KEY

CSC
CAPITAL SHOPPING CENTRES PLC

UXBRIDGE
SEPTEMBER 1998

LESLIE JONES Architects

issue registers; PowerPoint for presentation, some of which incorporate simple use of CDs. There is one copy of ClarisWorks which is hardly used. The firm also has copies of the NBS on disk complete with Specification Manager software for specification writing by the two specially appointed specification writers.

File Maker Pro software is available for databases with Claris Organizer used for contact lists, diaries, etc.; Microsoft Project for project planning although little use is made of the software to exploit resources scheduling capability. One copy each of QuarkXpress, Photoshop and Illustrator is available, retained since the dissolution of the graphics section, most of whose work is now outsourced with small amounts completed in-house.

The accounts section of the practice (with two members of staff) has its own separate IBM network linking directly with the bank and is not linked to the main practice network. Timesheets are done manually and collected weekly by the accounts section for input into their system.

IT applications for CAD

In 1995 the practice had three IBM PCs for secretarial machines and four Apple Macintoshes for the graphics department without any CAD except in the Manchester office with the Espirit CAD system.

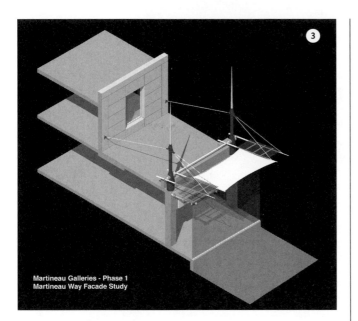

Martineau Galleries - Phase 1
Martineau Way Facade Study

The London office has about 54 PowerMacs (ranging from 8200–9600) purchased within the last two years. While all architects have CAD workstations in line with the policy of one machine per architect; all senior architects have 1400C Apple Powerbooks with remote access to the office, a facility which allows them to work from home and thereby avoid late working in the office. The standard CAD software MicroStation is used for larger projects (£10 million + contract value) with MiniCad used for smaller projects. Most 3D is to test elements and check that details work. Cafgen, the RIBA certification programme in Windows, is also available within the London office.

Typically the process within the practice involves carrying out development concepts on the remaining five drawing boards before doing the work (mostly 2D layouts and co-ordination) on screen. Each project except in the London office starts with a level structure based on the CISfB classification. Important drivers within the practice are client expectations of hand-drawn perspectives as well as CAD outputs.

IT applications for telecommunications and networks

Leslie Jones, London, has an Apple Macintosh network for all CAD workstations and five administration machines including one PC.

There are three servers — one for MicroStation work over £10 million value; the second for all the other CAD work and the third COM server handles all documents, E-mail using QuickMail, Internet access and dealing with documents management system and ISDN links with clients, printing company, design team members and so on.

Security is enhanced by password protection. Design team leaders have the responsibility of ensuring good communications throughout the project process. They do so by setting out a suitable IT basis and approach.

IT training

Two types of training are provided by the firm

(a)   for novices involving going through basics
(b)   conversion training employing an outside consultancy.

At least five training sessions are carried out a year. A policy of employing MicroStation literate staff is also pursued in order to improve the current 85% CAD fluency within the office. Vendor support is provided through Metaphor, suppliers of their equipment and software. Bentley Systems also provide MicroStation support including development of specific macros as necessary.

Possible developments for the practice include adopting ModelServer Publisher for which they have had a month's demonstration. ModelServer Publisher, a web-server-based publisher of engineering documents, allows anyone to dynamically view and query MicroStation design (DGN) files and other files, such as AutoCAD drawing (DWG) files, across a corporate intranet or the Internet. Viewing the published data requires only a standard web browser such as Netscape Navigator or Microsoft Internet Explorer. The testing of the product by the practice is consistent with the aim of adopting faster and easy web-based technology. For this purpose firewall protection from Metaphor is being investigated.

There are two issues of importance to the practice

(a)   the slow speed of the server posing as the firm's biggest problem
(b)   bandwidth.

The speed of the network and the amount of traffic and implications of legal aspects are general to the industry as a whole.

Selected projects

Uxbridge Retail Project

With a value of £65 million this project represents the first 100% CAD project for the office. Both MicroStation and MiniCAD allowed decision-makers within the practice an understanding of how CAD systems work. A lot was learnt and the project was crucial in the amount of hardware investment.

1. 2D layout drawing

Sunderland Project

2. 3D illustration of the £35 million project

Martin Perry emphasized that 'solutions must always respond to individual problems. Key areas of work will relate to improving tenant mix and rental return, improving the shopping environment and upgrading fire safety precautions.'

Central Birmingham Retail Centre

The £235 million project offers more opportunities of developing IT applications to a more sophisticated level than ever before especially in terms of projects implementation and how to meet the need to use reference files.

3. 3D CAD drawing forming output of the Façade Study

Central Stainer Project

4. This project (£28 million contract value) was for a multi-plex cinema, retailing and restaurants showing use of IT tools to produce a site photograph

Lifschutz Davidson, Architects | 16

Background

Lifschutz Davidson was formed in 1986 by two founding partners — Ian Davidson (formerly with Richard Rogers & Partners for three years and Fosters and Partners for five years) and Alex Lifschutz (formerly with Foster and Partners for six years). These, together with another partner, Paul Sandlands, form the three directors of the firm.

The practice has 24–25 staff, 20 of whom are architects, the remainder providing administration support. The firm has grown from a core group made up of directors to a first level of development, then to the second level of 10–12 persons and now a third level of 25–26 which is the optimum desired size. The firm has a policy of employing fewer and more experienced people.

Projects are commissioned through repeat business, word of month and competitions. Little promotional marketing material has been produced to-date. The range of projects is wide and varied consisting of bridges, urban regeneration, community housing, loft apartments, retail commercial and offices. Clients include J. Sainsbury, Harvey Nichols, Manchester Loft Corporation, restaurants, offices, corporate HQs, etc.

The firm's process is to define project objectives. The client acts as a catalyst for design ideas and innovation including alternate ways of satisfying the client's needs and requirements. The design

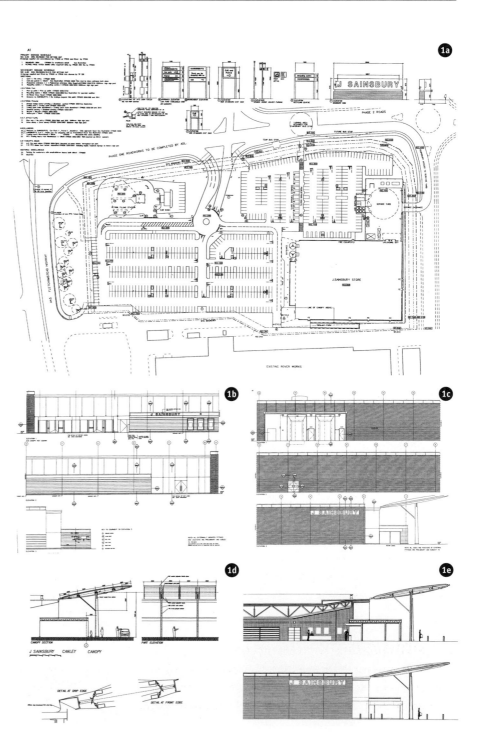

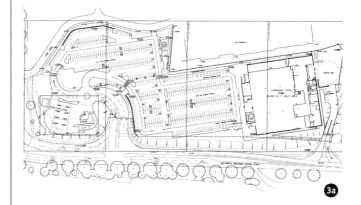

2

philosophy is 'maximum impact and quality'.

IT is important to the firm. The practice's IT strategy involves upgrading software/ hardware to meet project needs with finance as the main driver. Charles Borthwick is the CAD/IT manager usually supported on a day-to-day basis by other members of staff.

IT applications for text manipulation, databases and specification writing

Microsoft Office is used, in particular Word for text manipulation and word processing; Excel for spreadsheets, File Maker Pro for clients and contacts lists. There are four workstations for administration purposes.

Pegasus Software is used for accounts and is handled in-house by the office manager (Christine Otter) supervised and supported by the firm's accountant. Timesheets are done manually before being passed to the administration for resource programming and resource planning/allocation. This is based on working out the number of drawings required (man days) and aggregating these to produce monthly reports and the number of people required for each project.

Other software used by the practice includes QuarkXpress and Photoshop for brochures and reports. The NBS is used for writing specifications for every project.

IT applications for CAD

The practice has an ethernet network for CAD workstations, printers and plotters with administration workstations linked by Local talk. All computers are linked to an Apple Workgroup Server with mirrored

hard disks and DAT tape back-up facilities. There are six Apple Macs (Performas), fourteen Apple PowerMacs workstations, one Pentium PC and a range of printers, plotters and scanners including HP750C design jet and Epson colour ink-jet. More sophisticated printing for

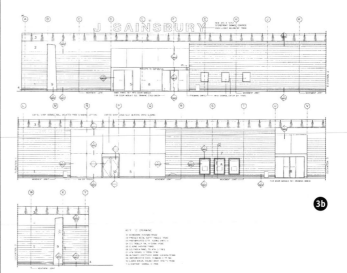

3a

3b

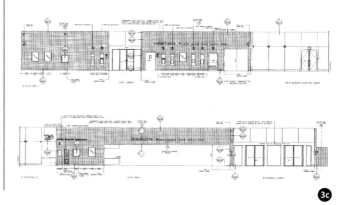

3c

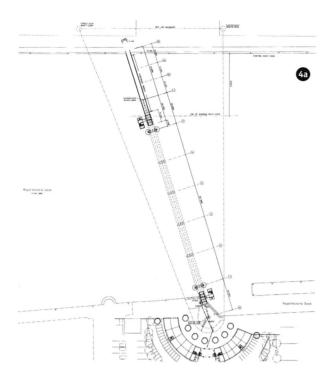

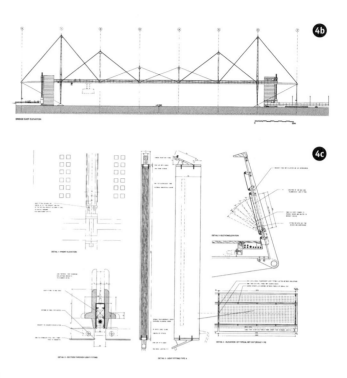

presentation purposes is carried out by bureau services.

MicroStation CAD software is used for 2D drafting and 3D modelling and rendering. The firm's aim is for the majority of architects to have their own personal workstations but with shared access to drawing boards for design and presentation purposes. An outworker carries out all the specialist 3D CAD work from home and is connected to the office through ISDN and modem for fast and efficient data transfer.

IT applications for telecommunications and networks

Other than the ethernet network the practice has an ISDN link allowing access to the Internet via CompuServe. This allows links and communication with clients and consultants, particularly transmission of documents, and assisting with project and marketing information. Since May 1998 the firm's new premises allow the practice to have a new telephone system, ISDN lines, voice mail, direct dial and CAT structured cabling.

IT training

The practice uses vendor support and troubleshooting. In-house training is the responsibility of Mandy Bates. The priority for the firm is to improve staff CAD skills and cross training new staff to MicroStation.

Selected projects

These illustrate how the practice exploits IT tools to translate design concepts into physical form and to create 2D and 3D drawings for its projects.

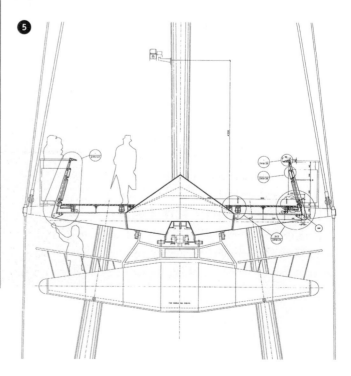

J Sainsbury Store, Canley, Coventry

This project is notable for its imposing front canopy to welcome the customer to the highly visible store. The 'wing' fascia of the Canley store is duplicated in the design of the on-site petrol station.

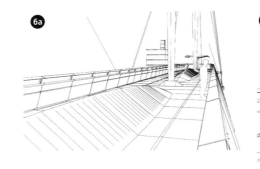

1. (a–e) 2D CAD drawings (plans, sections and elevations) include construction details (courtesy of J. Sainsbury)
2. 3D drawings include free-hand working details, e.g. cutaway view through external wall (courtesy of J. Sainsbury)

J Sainsbury Store, Garston, Watford

'A simple and elegant design that complements its important location on a strategic route into the town centre.'

3. (a–c) 2D CAD drawings (plans, sections and elevations) include construction details (courtesy of J. Sainsbury)

The Royal Victoria Dock Bridge, London

This new footbridge crosses the Royal Victoria Dock and is based on the century old principle of the 'Transporter Bridge'. Pedestrians can cross the dock by foot on a lightweight bridge deck accessed from pairs of scenic lifts and stair cores at each end or in the future, when all necessary approvals have been obtained, they can cross inside a 'people mover' cabin suspended under the bridge.

4. (a–c) The 2D CAD drawings (plans, sections and details) show the footbridge design and construction of mild steel with cast aluminium balusters and perforated stainless steel balustrading and bridge deck made of grooved iroko with black anti-slip resin inserts. The foot bridge links the West Silvertown area on the south side of the dock to the Custom House DLR and BR Stations in the north
5. The structural design is based on an inverted fink truss with six cable-stayed masts

providing support to the bridge deck which spans between 15 m high trestles. No part of the bridge deck is less than 15 m above water level and the bridge has a minimal cross-sectional area. Where the structure deepens, it extends above rather than below bridge deck level. This results in a series of distinctive humps on the bridge deck which are reminiscent of the keels of upturned boats. They can be used as vantage points to watch sailing events in the dock. In determining the design and height of the bridge, specific account was taken of extensive use of the dock for sailing and windsurfing. The design of the bridge and its elevated height responds to the dock's new use as a registered intersports centre

6. (a–c) CAD images were used extensively for visualizing the design (testing options) and presenting images for client and general consultation approvals. All construction information was produced on CAD in 2D and specifications utilized the NBS word processing supported package. Co-ordination with consultants, client body and adjoining developers was all helped by the exchange of digital information for setting out and general arrangement purposes

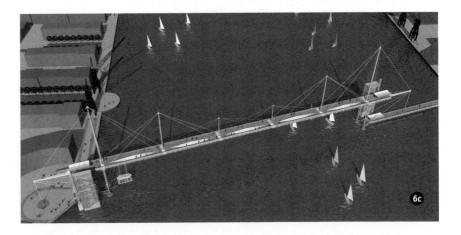

Michael Wilford & Partners, Architects	**17**

Background

James Stirling began the practice in 1956 and was joined by Michael Wilford in 1960. The Stirling/Wilford partnership was established in 1971 and continued until James Stirling's death in 1992. The directors Michael Wilford, Laurence Bain and Russell Bevington formed a new partnership in 1993 under the name of Michael Wilford & Partners which incorporated all rights and responsibilities of James Stirling, Michael Wilford and Associates.[1-3]

The firm is located in London, and has branch offices in Singapore and Stuttgart. It has an international reputation for producing buildings of the highest architectural quality which satisfy the requirements of the client's brief and respond to the opportunities of site and context. The firm is eminent as 'design' architects, fully involved in all phases and aspects of work with the professional goal of excellence, as outlined by Michael Wilford in the following statement:

Excellence requires integrity and control of the design process from concept through detailing and construction. For us design is an explicit and iterative process involving research, conceptual design, consideration of alternatives, their evolution and re-evaluation. We do not believe in waiting for 'the blinding flash' of inspiration. All aspirations and constraints are reviewed with the client and reconciled into a final project brief, which forms the basis of a thorough and wide-ranging diagrammatic exercise to establish all possible ways to configure the buildings. Alternatives are assessed and the range of options narrowed down and the concept diagram systematically developed to form a firm basis for the construction drawings and specifications.

The practice workload includes the design of buildings for cultural/institutional clients: museums, universities, central and local government, new town corporations, United Nations; for corporate clients such as Olivetti, Siemens, Bayer, B. Braun Melsungen, City Acre Property Investment Trust, Chelsfield and Olympia and York, and individual patrons of international renown, for instance Lord Palumbo, Baron Thyssen and Rina Brion.

Buildings and projects by the firm have been illustrated extensively in publications and included in exhibitions throughout the world.[4-6] Major exhibitions tracing the evolution of James Stirling, Michael Wilford and Associates into the current practice of Michael Wilford and Partners were held in London in 1996, in Edinburgh and Barcelona in 1997. The work by the practice has also been the subject of TV and film documentaries. James Stirling was the recipient of many awards and Honorary Fellowships, for instance, the RIBA Gold Medal in 1980, the Pritzker Prize in 1981 and the Praemium Imperile Award in 1990. In 1997, the practice was awarded the Sunday Times/RIBA Stirling Prize.

The principals are based in the London office and share responsibility for all projects. An associate is responsible for the day-to-day

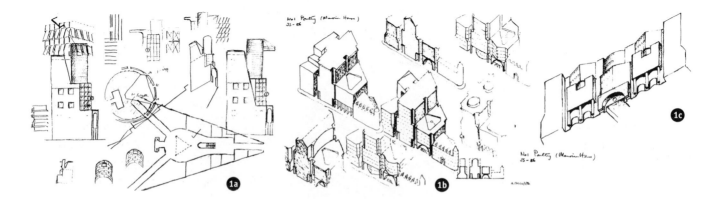

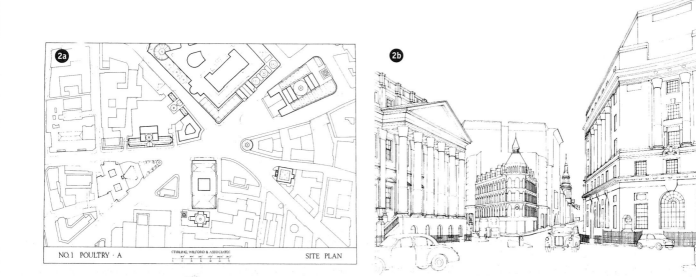

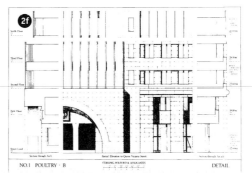

NO.1 POULTRY · A   SITE PLAN

NO.1 POULTRY · B   STIRLING WILFORD & ASSOCIATES   DETAIL.

aspects
of organiza-
tion, programming
and production of each
project. All staff (over 30) are qualified
architects. There are no other in-house
disciplines and preference is given to assembling
specific and appropriate teams of consultants for each
project. Projects employ a variety of procurement methods
according to client needs, programming, budget, site
constraints and construction methods. Guidelines are laid
down on a project-by-project basis in order that the

practice's approach and
procedures are best suited to the
country of origin of the project
and its approval system.
Projects in North America and in the Far
East are executed in association
with established local
firms. Typically mas-
terplanning, schematic
design and design
development work are done
in the London office with assistance

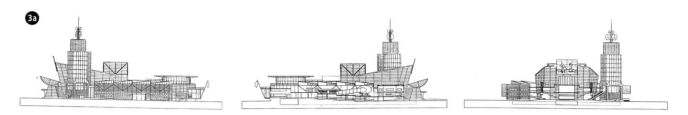

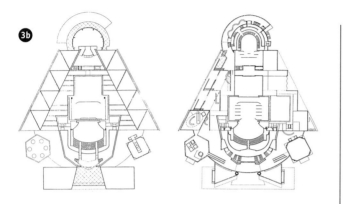

from the local architect providing input on climate, codes of practice and regulations. Working drawings, construction documents and construction supervision are carried out by the local architect with London providing continuity of design intent through until project completion.

Quality control is maintained based on previous experience with each project directed by a partner, associate and project architect all of whom are involved from inception to completion and oversee the design, development and all subsequent stages of the project. Senior personnel authorize operating procedures for constructive feedback, ensure that all necessary information is available to project team members and continuously reflects any changes to the client's brief and is co-ordinated with the work of consultants.

All procedures and work are reviewed continuously to ensure that the design, the progress and quality of work, the receipt and transmission of information meet design objectives, the programme and fulfil both client and office aspirations.

Since the introduction of CAD, IT tools (hardware and software) have acquired an importance to the practice and responsibilities are shared by all users. There is no IT manager.

IT applications for text manipulation, databases and specification writing

The firm runs Microsoft Word, Excel spreadsheet, PowerPoint and File Maker Pro database software. Word is used for word processing as well as for specification writing, for example in the Tate in the North project. QuarkXpress and Adobe Photoshop are also available and are used to produce marketing information. All the practice software has full systems information transfer compatibility.

Timesheets are completed manually and passed to the accounts section.

IT applications for CAD

The practice computer system provides clients with state of the art computer-aided design helping to visualize and present designs in the most rapid and accurate way. The system is based on Apple Macintosh Power PC computers running on an ethernet network using Bentley MicroStation software with access to Modeler/MasterPiece and Triforma. Bentley MicroStation software

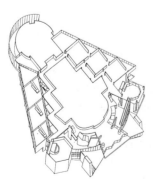

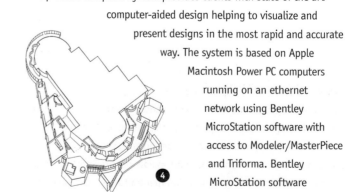

operating on an Apple Macintosh platform is fully compatible with AutoCAD drawing format operating on a PC platform.

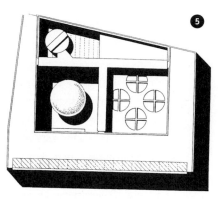

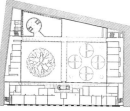

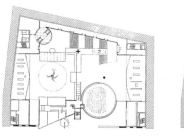

Within the practice the computer system is used for 2D and 3D design and coordination, 3D modelling, visualization and rendering. This aids

the development of a richer, inclusive architectural language based on the fusion of modernity and tradition and the thorough review and assessment of the schematic design by the client and other participants allowing a comprehensive investigation of all aspects including materials and colour.

For production of specific computer images the firm outsources to practices such as Hayes Davidson.

IT applications for telecommunications and networks

The firm is connected to 4Sight ISDN as well as Demon with full E-mail capability; technologies which have contributed to the extensive experience of collaborative working.

IT training

Newly employed staff receive initial training which forms part of the initial induction course. Experienced users within the office offer help-desk support and share their experience during project reviews and discussions. Full system support is provided by Ian Martin Associates.

Selected projects

In this section several projects were considered for selection, including The Tate Gallery in the North (Phase 2) in Liverpool, 5–7 Carlton Gardens in London (from a limited competition), Braun (Phase 2) in Melsungen, Germany, Salford Art Centre in Salford, The British Embassy in Berlin and Number 1 Poultry Mansion House Development, London.

The Tate in the North (Phase 2) is important in showing the strong emphasis and use within the office of the conventional or traditional approach and techniques for producing outputs such as 'worm views and axonometrics', historically identified with the practice statement. Drawings are done by hand, as with the original drawings, and facilitated by the small project team. There are no client requirements for use of CAD. The Carlton Gardens is important in another sense. It shares the approach developed for the Lowry Centre, and reacts to a tight timetable for the commercial development, the repeat elements within the design, and the enlarged size of the project team. The project for Braun (Phase 2) shows parallel working arrangements. Design information, incorporating 2D layout plans, is developed and communicated between London and Stuttgart offices via ISDN link. The Salford Arts Centre is more significant in that a conscious decision was made to operate within a CAD framework alongside physical models and hand-produced axonometrics (difficult to produce using CAD). The 3D environments are exploited to develop and rationalize specific elements, for example the 1650-seat Lyric Theatre, 400-seat Flexible Theatre and organic plans. The project for the British Embassy shares the approach initiated with the Salford Arts Centre.

Three projects, Poultry 1 Mansion House Development, Salford Arts Centre and the British Embassy, are illustrated.

Number 1 Poultry Mansion House Development, London

1. (a–c) Plans, sections, elevations and details were produced for a variety of schemes such as Schemes A and B and for different presentation purposes including the two Public Inquiries on the development.[7] The sketches are done free-hand. In general the free-hand sketches represent the conceptual design stage, and design development studies are only now being addressed efficiently by intuitive conceptual design software and advanced graphic communication tools such as Adobe Illustrator, which

are still to gain wide acceptance within many practices

2. (a–h) 3D illustrations including perspectives, axonometrics, etc. There is a significant gap between the level of skill and sophistication of the hand- and computer-produced drawings. The skill and expertise required in producing the axonometrics and the different presentation drawings poses a tremendous challenge to both the computer software developers and hardware manufacturers as well as the associated computer users' training requirements and expertise

Salford Arts Centre/multi-purpose complexes, Salford [8]

3. (a–b) 2D layouts incorporating plans, sections and elevations indicate the space planning to house: a children's gallery, study centre, virtual reality centre, Lowry gallery, and two performance areas — the 1650-seat Lyric Theatre and the 400-seat Flexible Theatre

4. 3D illustrations dominated by axonometrics and also including physical models and the spirit of the Arts Centre Complexes which is designed to act as a catalyst for further, mixed development and the revitalization of Salford

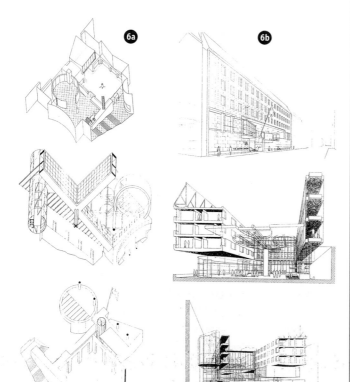

British Embassy, Berlin, Germany [9]

5. 2D layouts show various locations, site plans incorporating ordinance survey maps and floor plans (including layered floor plans of the building), sections and elevations. The design revolves around a series of grand spaces. Visitors walk through the entrance court and up the main staircase to the piano nobile. The upper landing of the stair marks a transition between the entrance hall and the raised winter gardens. Offices and balconies are organized to encircle the upper levels of the central winter garden with the chancery located at the front of the building

6. (a–b) 3D illustrations are indicated mainly by sophisticated and clearly identifiable cutaway axonometrics from the practice and by perspectives. The sectional perspectives highlight the winter garden, the internal focus of the embassy and the building's stone façade matching the stone of the Brandenburg Gate in the historic location. The façade shows the three zones — base, ceremonial route and offices — forming a layered expression of the building's internal organization. Elements such as the ambassador's offices are allowed to break through the façade. On ascending the final flight of stairs into the grand space, the public rooms of the embassy — the circular conference room and dining room on the axonometric — gradually unfold. All these images include physical models

References

1. Rattenbury K. Establishing the succession. *Building Design*. No. 1272, 12 July, 1996, 2.

2. Morre R. The genius's apprentice: Michael has emerged from the shadow of James Stirling. *Blueprint*. No. 129, June 1996, 40–43.

3. Stirling J. and Wilford M. James Stirling, Michael Wilford and Associates: an architectural design profile. *Architectural Design*, **60** (5–6), 1990, special issue.

4. Stirling J. and Wilford M. *James Stirling, Buildings and Projects*, Rizzoli, New York, 1984.

5. Michael Wilford and Partners. *Wilford Stirling Wilford*. Published by Michael Wilford and Partners to coincide with the exhibition 'Wilford Stirling Wilford' at the RIBA Architecture Centre, 10 June to 3 August 1996.

6. Wilford M. Our kind of architecture. *Architects Journal*, **203**, No. 23, 3 June, 1996, 32–33.

7. Wilson, Colin St John *et al*. Poultry No. 1 in London: Architects James Stirling, Michael Wilford and Associates and Ove Arup and Partners. *Zodiac*, No. 3, 1990, 62–67.

8. Baillieu A. Salford reveals designs for £75 million performance arts centre. *Architects Journal*, **201**, No. 13, 30 March, 1995, 12.

9. Finch P. Berlin game: winning scheme by Michael Wilford and Partners. *Architects Journal*, **201**, No. 9, 2 March, 1995, 8–9.

Mott MacDonald, Consulting Engineers | 18

Background

Mott MacDonald is a multi-disciplinary engineering consultancy, a wholly independent UK-based company with 4500 staff worldwide, turnover approaching £200 million and operating in over 100 countries.[1,2] According to *Engineering News Record* the firm is Britain's top exporter of engineering consultancy services, based on the 1996 earnings of international firms.

The structure of the firm (Fig. 1) shows that Mott MacDonald Group Ltd comprises central services, technical and regional management units and companies. Technical management units are made up of Civil & Transportation, Industrial & Commercial, Water & Environment, Management Services and Power and Communications. Regional management units consist of Mott MacDonald UK, Mott MacDonald International, Mott Connell Ltd (Hong Kong), Hatch Mott MacDonald Inc. (North America) and Connell Wagner Pty Ltd (Australia & New Zealand) — all are regional companies.

Landmark projects include the Channel Tunnel, Hong Kong's new international airport passenger terminal and record-breaking Lantau Link, London's Heathrow Express and Twickenham Rugby Ground redevelopment, Los Angeles Metro Extension, Lesotho Highlands Water project and Shanghai Environment project. Landmarks in the power sector include China's first private power station and the second phase of Saudi Arabia's Shoiaba power and water station featuring the largest single desalination contract ever placed. UK projects vary in scale and complexity covering various contractual arrangements involving work on BOT schemes, on term contracts, as part of a client's own team or for contractors on fast-track design-and-build packages.

The firm's technical base covers the full spectrum of engineering disciplines allowing it to provide a totally integrated and quality assured one-stop service covering all aspects of civil engineering, transportation, power and energy, building and infrastructure, water supply and waste water, mechanical and electrical engineering, environmental management and rural development. Capabilities include offshore engineering, communications and control systems, geotechnics, hazard and risk assessment, information technology and project management.

The firm employs engineers, environmentalists, planners, scientists, economists, architects, project managers and computer specialists. Every commission is assigned a project director — with overall responsibility for it — plus a project manager who is the client's principal contact and who oversees the project's day-to-day management and cost control. Close teamwork with clients ensures that all project activities are fully co-ordinated and targeted towards their objectives with special training provided for clients' staff if required.

Information Technology is important for the firm which has an annual budget for this purpose. The strategy is to move towards more qualified engineers (on a one PC per person basis) and fewer technicians/draftsmen. John Gill, a transportation engineer by training is the director for IT services and his responsibility covers the group standards software chosen by the central unit.

The firm operates a quality system in accordance with British and ISO standards to guarantee quality of service.

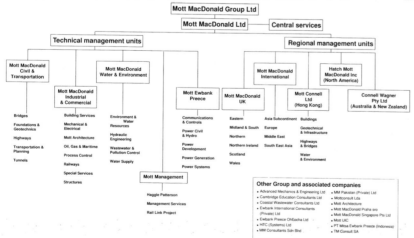

Fig. 1. Group management structure

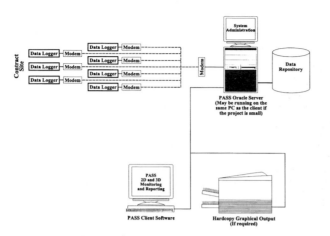

Fig. 2. PASS 3D software package

IT applications for text manipulation, databases and specification writing

Group standards software based on Microsoft Office (Word, Excel, Access) is used for administration purposes, spread sheets, small databases and project planning. Support and help-desk answers to queries are provided to users throughout the organization. Different sections may adopt different software and hardware to suit specific needs.

IT applications for engineering analysis and CAD

Technical software for design, CAD and analysis (as part of the internal practices) is the responsibility of the different sections of the group. For example, the bridge design section has a range of software from structural analysis, finite analysis system software to

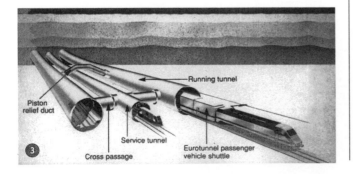

CAD. Nonetheless, AutoCAD is used for 80–90% of the project work since upgrading from the GDS system. Bentley's MicroStation is also available to meet specific needs.

The size of the group and the range of services offered inevitably means that there are many software packages used throughout the organization. The important issue for the firm is 'leading the market and in particular relating to the client, meeting client requirements and making acceptable charges to the client to ensure profitability'.

The practice has developed several in-house software products — BRIGIT a semi-automated bridge design programme expanded to handle earthquakes as well as typhoons; a computer integrated road design package, and commercial software covering all aspects from preliminary design to bill of quantities. Development of powerful new computer animation and virtual reality tools allows the firm to dramatically increase the comprehension of very complex computational fluid dynamics (CFD) simulations, especially in smoke behaviour modelling where smoke layers can be visualized in a virtual world directly imported from the CFD model.

PASS, an in-house 3D software package (Fig. 2) allows prediction and analysis of structural settlement associated with tunnelling in urban areas and thus enables safeguards to be made against building damage. Other developments include a mathematical and graphical technique for on-site control of pipe jack alignment and a suite of programs claiming increases in efficiency, accuracy and speed of up to 40% for drainage area studies.

Integration of the different technical and group standards software while preserving local practices and experimentation is an important issue for the practice.

IT applications for telecommunications and networks

The size of the group means communication is crucial, both internally, within the group and externally with clients and other consultants as members of design and project teams, especially since two-thirds of the business is now for overseas clients and only one-third in the UK. The enterprise network with worldwide connections covers the whole of the UK linking each of the office networks (mostly Novell with Windows NT in about 50% of cases). The worldwide links are dominated by E-mail. The location of the web site is *mottmadonald@compuserve.com*.

IT training

Mott MacDonald is committed to providing training and management skills to sustain the benefits of investment for future generations. There is a budget and there are scheduled training courses using in-house as well as external trainers. Divisional managers monitor skills within their division and allocate and recommend training to individuals who can also apply for special training. They also have to balance the cost of training and other needs of the division.

The practice also makes staff available to voluntary organizations such as Save the Children and RedR (the UK Register of Engineers for Disaster Relief).

Selected projects

Such projects show the way IT tools are applied in the firm to 'develop imaginative, innovative solutions for demanding technical problems to the benefit of clients and the company'.

Hardcopies are made of project work to meet legal requirements and as back-up. The hard copies are transferred on to microfilms for archiving purposes. This has proven valuable in projects which have been revisited including some which were translated from the GDS system to the AutoCAD system.

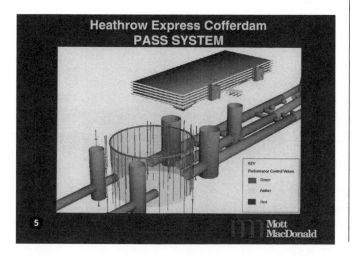

The Channel Tunnel linking Folkestone, England and Sangatte, France

This is a project, for Transmanche Link — TML — for Eurotunnel (£1.3 billion at 1993 prices), in which Mott MacDonald provided preliminary and detailed design of civil, mechanical and electrical works on the UK sector. The firm's involvement in schemes to build a tunnel under the English Channel goes back to a feasibility study undertaken in 1929.

Mott MacDonald's design of the 50 km tunnel, and subsequent investigations and tests in 1986–87 covered aerodynamics and ventilation, alignment, drainage, track form, tunnel lining and geotechnics. Computer simulation of airflow around shuttle trains was part of the development.

3. Schematic cross-section of the Channel Tunnel. The UK cross-over cavern — the world's largest subsea excavation at 156 m long by 18 m wide and 10 m high internally — was constructed using the New Austrian Tunnelling Method. Movement of the cavern excavation was monitored by means of extensive instru-mentation at some 200 stations

Heathrow Express Rail Link for BAA plc

For this project (£235 million at 1989 prices) Mott MacDonald provided lead design services for the project and expanded by commissions to design the trackwork and all mechanical and electrical installations (apart from signalling) including ventilation and electrical installations. The high-speed link between Heathrow Airport and London's Paddington Station has been designed to provide specifically commissioned trains every 15 minutes in both directions with an estimated journey time of 16 minutes.

4. The scheme includes 10 km of tunnels and two new stations, one serving Terminals 1 and 2 and the central terminal area of Terminal 3, the other serving Terminal 4

5. Extensive use was made of the in-house software packages PASS and DATAMAP in the construction of Heathrow Airport's 60 m diameter cofferdam which encompassed most of the disturbed ground that resulted from the collapse in October 1994, when sections of the station tunnels collapsed during construction thereby creating a significant setback for the completion of the programme. The cofferdam allowed safe excavation and created unimpeded space for bottom-up construction of the station.[3]

Instrumentation and monitoring formed an integral part of the overall risk management of the cofferdam.[4-6] The principal objective of the instrumentation was to monitor the response of the ground and the structure to construction so that design assumptions could be reviewed and validated and the actual behaviour of the structure demonstrated. In order to concentrate on key design/monitoring parameters, the instrumentation system was split into primary and secondary

monitoring systems. The primary instrumentation, which comprised inclinometers and precise levelling, was monitored at frequent intervals and formed the basis for decision making on pre-designed contingency actions. Secondary monitoring systems were used to provide back-up to primary systems and to give global/background information.

The instrumentation included 744 beam-mounted electrolevel sensors both in piles and in the ground, 71 extensometers, 2442 strain gauges, 33 piezometers, 407 prisms and 105 survey points. Mott MacDonald design teams led the monitoring and interpretation on site and identified the actions necessitated by the findings of their interpretations. Remote data from instruments located in and around the cofferdam were captured through data loggers, modems and computers. The data were read every hour and stored in the computer. The system was capable of higher data collection rates (if required), interrogation by model links or by a directly-linked computer, in case the telephone lines failed. The raw data, including survey readings, were processed using the data processing software DATAMAP. This software also enabled relevant processed data to be accessed and plotted in the required form.

Lantau Link, Hong Kong

This project consisting of two bridges, the 1377 m main span Tsing Ma suspension bridge (**Fig. 6**) and the 430 m main span KapShui Mun cable-stayed bridge (**Fig. 7**), form a strategic part of the road and rail connections between Hong Kong Island and the new airport at Chek Lap Kok. Mott MacDonald as the Hong Kong Highways Department's engineer was responsible for the planning, design, construction contract management and completion within budget in May 1997 of the Lantau Link.

The design challenges, linking stringent geotechnical and operational requirements of a 135 km/h railway crossing, governed the layout of the bridge. The air corridor for the new airport imposed restrictions on the 206 m tower height and ocean going ships passing beneath demanded a headroom clearance of over 60 m. In addition, the bridge has to remain serviceable during the region's very severe tropical storms.

8. A streamlined vented girder was developed, wind-tunnel tested and used to achieve the required strong and stable deck.

Streamlining was used to minimize drag and venting, and to enhance stability. The extremely compact innovative deck cross-section measures just 41 m wide by 7.7 m deep and called for around 50 000 tons of steelworks. After fabrication, suspended span units — mostly 36 m long and weighing over 1000 tons — were assembled in a Chinese yard on the Pearl River delta, floated to site on barges and hoisted into position

Careful attention to aesthetics, the proportions, shape and finishes of the towers was an integral part of the bridge's design which aimed for overall elegance and harmony.

References

1. Mott MacDonald Group Annual Review 1996/7.

2. Newsletter of the Mott MacDonald Group. *Momentum,* Issue 17, winter 97/98.

3. *Ground Engineering,* Sept., 1996, 18.

4. Institution of Civil Engineers. *The observational method in geotechnical engineering.* Thomas Telford Ltd, 1996.

5. *The observational method.* Engineers Educational Channel, TEN, ICE, May/June, 1996. Video.

6. Powderham A. J. An overview of the observational method: development in cut and cover and bored tunnelling projects. *Géotechnique,* **44**, No. 4, 1994, 619–636.

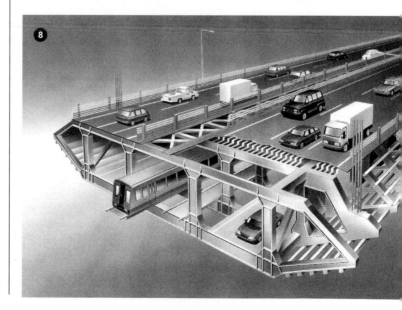

Oscar Faber, Consulting Engineers | 19

Background

Oscar Faber is a multi-disciplinary engineering practice founded in 1921 which provides a complete range of engineering solutions for all aspects of building engineering and transportation planning. The group is organized according to self-managing (mostly specialist) business units each of which is led by a Director answerable to the Board — building services (mechanical & electrical); structural engineering; transportation (planning); Kuala Lumpa office; acoustics; facilities management; geotechnology; environmental group; infestation; research & development. The first two groups form the core business of Oscar Faber, Consulting Engineers and represents 40% of all the work carried out by the group. The total number of staff is 200 in the St Albans Head office, and 600 UK wide.

Work comes from two main sources — repeat business and a result of marketing exercises. The old routine of architect and contractor relationships obtaining work through contacts has been largely replaced by EC procedures for projects of a certain size and complexity. The range of work covers offices (such as Canary Wharf), leisure, prisons, railway buildings, housing, laboratories (Fisons plc, pharmaceutical division). Most of the projects in which Oscar Faber are involved are organized either traditionally around an architect or around a consortium leader, who might be an architect. Clients come from all different sectors, e.g. British Telecom, Royal Bank of Scotland, TSB, Norwich Union, Bank of England, Bank of America, IBM, Woolwich, Pearson, Grand Met, Barclays De Zoete Wedd, Lloyd's, Motorola and Marks & Spencer. The projects are developed in a teamwork situation with systems (including compatible computer systems) pulled in as necessary to support the project and the team. For example the Birmingham Children's Hospital used Sonata CAD system. The extent and the nature of work is determined and allowed for within the fee estimate.

Traditionally, the process has been that the architect produces drawings which are then issued in paper copies to the engineer who then develops the engineering content using the services of a technician (a tracer, etc.). For a time the technician was replaced by a CAD operator. The objective and the practice at Oscar Faber is to make every engineer CAD fluent including carrying out administrative tasks such as typing letters and minutes directly using macros, thus reducing the need for secretaries. Also the Quality Assurance procedures to ISO 9001 are of importance to the way in which projects are structured and carried out. These entail monthly reviews and internal audits on an *ad hoc* basis for each project as well as external audits (one per year) to satisfy the accreditation requirements.

Within Oscar Faber each project is organized around an assistant director who reports to the Board directors and co-ordinates the design. Below him are principal engineers (mechanical, electrical or other), the number of which will depend on the size and complexity of the project. Oscar Faber have an in-house IT group which develops its own software and provides consultancy services on developing, resourcing, managing and operating IT within the firm. The group can offer continuing IT training to any member of the Oscar Faber group.

Dr A. G. Selman leads a division in the practice as a specialist with 25 years' building services experience of the health sector (five years with Oscar Faber). He is also the resident PFI (Public Finance Initiative) specialist. He acknowledges that IT is important. Its integration within the design supply, construction and management of completed buildings can affect both the process as well as the product. Barriers to this are the culture of doctors and ward clerks resistant to change and to new methods of working. This includes dated plug-in philosophy of 'Building Notes' which is no longer relevant. The change is due to the need for income-generating activities, PFI and its emphasis on affordability, and is coming from the US where there is a greater awareness of the value of using IT. The equivalent ward clerk registers a patient and IT makes it possible to record his/her requirements (order food, diet, surgical kit, instruments, etc.).

IT is a useful 'medical tool' important in situations where the building is no longer considered a 'free good' whose asset is not allowed to depreciate. The size of a building is related to both capital costs as well as recurrent costs (heating, cooling, lighting, ventilation, etc.) Larger hospital buildings cost more to build and are expensive to run. IT can aid reduction in building size and construction costs thereby making achievement of optimum life-cycle costs possible. IT can also help reduce the average stay in hospital, currently at ten days (target of 3.5 days by 2020). All this indicates a pressing need for a mature robust IT strategy with enough system redundancies (whether software or hardware) to meet varying operational needs.

IT applications for text manipulation, spreadsheets and databases

WordPerfect is the preferred word processing package for the practice which also uses Microsoft Office without Word; Excel for spreadsheet, PowerPoint for presentations and the practice's own software for some of the activities.

Oscar Faber have an enormous database in Word incorporating the NS specification. For every project specification writers work and modify the database as necessary to produce the specification for the project.

IT applications for CAD

In the past the group has had an in-house IT section developing its own programs, but this was sold-off in order 'to concentrate on core business'. Off-the-shelf packages are preferred to meet project needs. The practice uses AutoCAD and MicroStation with MicroStation as the preferred practice standard. The group uses many software packages such as Hevacomp for M & E depending on the discipline and division.

Oscar Faber has a relationship with Centre for Industrial & Medical Informatics (CIMI) based at the University of Nottingham. The centre specializes in advanced computer imaging technologies. The heart of its operation is the Reality Centre, a state of the art facility for real time, human scale computer modelling and virtual environment. Clients can 'fly' through virtual models at full scale and assess environmental changes at an early stage. The facility can

also be used to explore 'what if' scenarios and undertake investigations that otherwise would be uneconomical or impossible.

The Bluewater Project, Kent, involving Bovis is an example of projects which utilized the facilities at the centre. The project is also important in using a computer management system sponsored by Bovis as a project management group. 'Humming bird servers' act as a repository of project data for ten or more design disciplines who use specific security and protocol to access the project information.

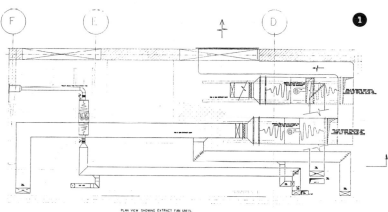

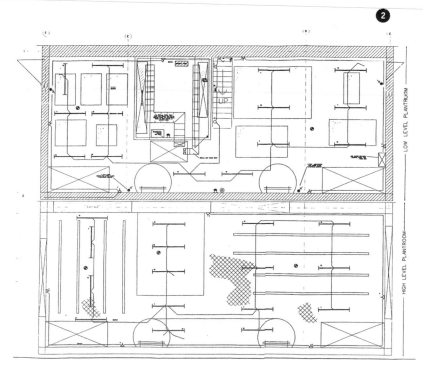

IT applications for telecommunications and networks

Oscar Faber operates a network of IBM PCs (386s, 486s and Pentiums) with two file servers. The St Albans network is linked to the London and Bristol offices. Links with other offices allow E-mail of graphics from Dallas where they are five hours behind the UK in time. The network is constantly developing to meet new client and project demands for more dedicated IT links.

IT training

The in-house IT section offers continuing IT training to any member of the Oscar Faber Group.

Other Issues

The practice operates a strict control of IT activities (anti-virus checks, etc.) following a burglary in which micro-chips were stolen from PCs causing great inconvenience. Computers are useful in addressing incompatibilities between the information supplied by different disciplines in a project. There is emphasis on ensuring that information input into the computer is carried out correctly.

Selected projects

The projects reviewed for selection are from the Healthcare buildings sector: high technology buildings in which over 46% of the cost may be attributed to Mechanical and Electrical systems alone. They include St George's Health Centre in London, Worcester Trust Hospital, Halifax Hospital and Birmingham Children's Hospital.

The Worcester Trust Hospital project (with RTKL architects, Dallas) comprises 360 beds and represents a rationalization of 6 existing hospitals. In this case the number of procedures is the most critical factor rather than the number of beds available (this is the trigger for change). For the practice, IT is also relevant for the Halifax Hospital project with RTKL architects (£76 million contract). The Birmingham Children's Hospital project is of further importance because it has been recently completed and has

received many good reviews.

2D drawings (**Fig. 1–4**) from the St George's Health Centre are illustrated here. These indicate the plantroom layout of electrical services and co-ordination of services. The drawings represent output from an emphasis of using analytical tools and manual skills at the expense of CAD methods and expertise. They also relate largely to the seminal work on heating and air-conditioning first published in 1936 by Oscar Faber & Co. and now in its eighth edition.[1]

Reference

1. Martin P. L. and Oughton D. R. *Heating and air-conditioning of buildings*. Faber & Co., 8th edition, 1995.

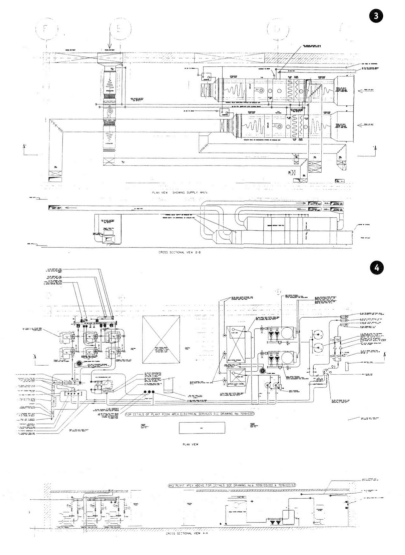

Price & Myers, Consulting Structural and Civil Engineers | 20

Background

Price & Myers was formed by Sam Price (18 years with Ove Arup & Partners) and Robert Myers (14 years with Ove Arup & Partners and Arup Associates) in 1978. Robert Afia (15 years with Ove Arup & Partners and Arup Associates), Nick Hannika (6 years with Ove Arup & Partners and Arup Associates) and Mike Lovick (5 years with Richard Costain Ltd and 9 years with Arup Associates) became partners in 1981, Steve Wickham (5 years with Ove Arup & Partners) in 1993 and Jonathan Darnell (with 5 years' experience including 2 years with French–Kier) in 1997.

The practice has grown to 50 staff in two offices in London and Nottingham. Work varies from minor alterations and extensions to existing buildings, to major new build and refurbishment projects. The work covers offices, public buildings, university buildings, industrial, repairs and restoration of historic buildings and civil engineering work. Clients include Gillette UK Ltd, H J Heinz, Alliance & Leicester Building Society, Colleges at Oxford, Glaxo, Crown Estates, BT, local authorities, housing associations and private developers.

A number of completed projects have received awards (Civic Trust, RIBA Regional, National and Energy Efficiency, Structural Steel Design, Concrete Society and Housing Design Awards)

All partners are actively involved in the design of projects. The practice carries out regular reviews of all major projects in order to share experience of working in multi-disciplinary teams and to foster good working relationships. Typically within the practice, engineers produce sketches and analysis while the draftsmen produce drawings from engineer's concepts and sketches.

The practice profile notes that computer-aided design and analysis are extensively used throughout the office and computer resource facilities are said to be consistent with a practice of their size and range of work. The partners and associates regularly review the IT requirements of the practice for which Robert Myers is responsible. Richard Crawford, a structural engineer by training, is the IT manager. He spends a quarter of his time implementing changes in the day-to-day operations.

IT applications for text manipulation, databases and specification writing

The practice uses Coral Suite as an office standard: WordPerfect (on four stations) for word processing — letters, memos, minutes, specification writing. Microsoft Word and Excel are also available.

Lotus 1 2 3 is the main spreadsheet and is used for accounts. The package allows importing of timesheet data from an in-house

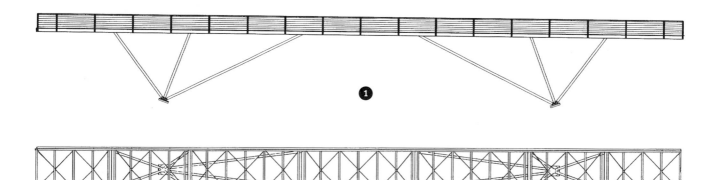

programme Job Man, a time recording programme maintained as a database of information on clients, contractors and consultants.

Microsoft Access has been used to create a list of jobs. Pagemaker, Paintshop Pro and HP DeskScan with 35 mm film scanner and flat bed scanner are available for making brochures and marketing literature.

IT applications for engineering

Ninety per cent of the firm's calculations are done by hand. The varied nature of the work done by the practice makes the use of computers cumbersome due to lack of flexibility of calculation packages. The human brain is quicker, intelligent and more selective.

Robert Myers

The more complicated structures — for example, staircases or complicated roof structures — are checked using Superstress, a package developed in the 1960s. The computer packages are also very useful in dealing with multiple load cases allowing a refined structure to be produced.

Software packages include Oasys AP+, OasysCompos, SCALE (an analytical structural engineering programme), and SuperStress which uses relational method and input produced by Integer SC Beam and SC Rcol. SuperStress is popular with the practice because it allows more graphical manipulation and output. Staad III which uses finite element analysis is available on one machine.

The firm is developing its own ability to produce 3D drawings. It uses AutoCAD Mechanical Desktop and Bentley's MicroStation. Two draftsmen have received the necessary training.

Most of the practice's software and hardware is supplied by Barclay Technical Graphics Ltd who also provide vendor support. All hardware (about 17 computers altogether) comprise either IBM 486s, Pentiums or IBM compatibles.

IT applications for telecommunications and networks

The firm has a policy that each computer is a self-contained unit 'to safeguard against the problems sometimes caused by networks or the reliance on a central server'. However, the CAD stations are linked pier-to-pier to allow the draftsmen to exchange drawing files and to use a common plotter.

The firm uses E-mail in the usual way and has found it particularly useful for sending CAD files to sites and other consultants. Price & Myers Internet location is *pricemyers@AOL.com*.

IT training

Price & Myers have a policy of encouraging the sharing of information and dissemination of ideas at lunch-time meetings. This includes any novel ideas and any knowledge acquired by experienced

users. The practice Friday meetings include a commentary on IT and use of computers.

The main issues for the practice concern sorting out the network, developing conformity of IT tools and hardware to overcome the legacy problems while not wishing to drastically change so as to alienate the staff and the way of doing things. The policy is more to be on the wings of technology.

Selected projects

A number of projects were reviewed for purposes of selection: a private house for Lord Sainsbury in France, the Torus Project in Wales (Architects Foster & Partners), COM Warehouse Project in Hemel Hempstead, Alliance & Leicester Building Project and the development for Computacenter at Hartfield Business Park.

The private house project for Sainsbury is the first time the practice was able to communicate electronically (sending and receiving drawings) for a site. The Torus Project (Architects Foster & Partners) is another first for the practice in that the complex geometry was handled through AutoCAD and analysis was carried out using spreadsheets. The 2D drawings, including the section of Torus, were facilitated by use of CAD. Without computers as a design tool, the complex geometry would have been difficult and lengthy. The enormous physical size of COM Warehouse is characteristic of projects for a US

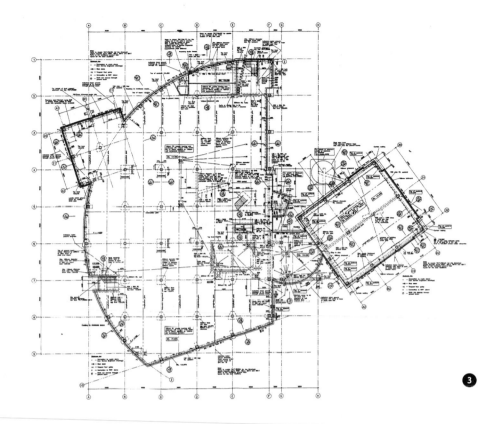

3

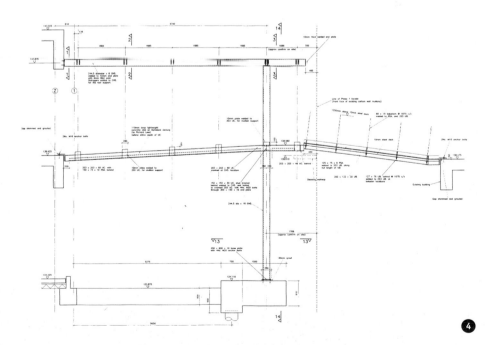

4

computer hardware manufacturer. IT tools facilitated analysis and design of the many repetitive elements and components.

Private house project for Sainsbury, France

1. Plan and section of the bridge
2. 3D illustration incorporating photographic information

COM Warehouse Project, Hemel Hempstead

3. 2D plans show the very large size of the foundation layout
4. Details of the link bridge — these contrast with 3D CAD stair details and studies using glass materials in **Fig. 5**

Computacenter, Hartfield Business Park

6. The 2D plan again indicates the very large size of the foundation layout (width of 26 structural bays at 7.5 m centres against 18 bays at 8.0 m centres)

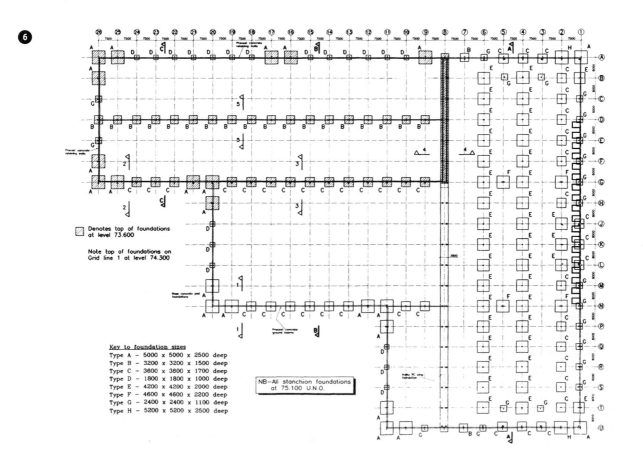

Denotes top of foundations at level 73.600

Note top of foundations on Grid line 1 at level 74.300

NB—All stanchion foundations at 75.100 U.N.O.

Key to foundation sizes
Type A – 5000 x 5000 x 2500 deep
Type B – 3200 x 3200 x 1500 deep
Type C – 3600 x 3600 x 1700 deep
Type D – 1800 x 1800 x 1000 deep
Type E – 4200 x 4200 x 2000 deep
Type F – 4600 x 4600 x 2200 deep
Type G – 2400 x 2400 x 1100 deep
Type H – 5200 x 5200 x 2500 deep

Reiach & Hall, Architects | **21**

Background

Reiach & Hall Architects (R&H) is a comparatively large Scottish practice based in Edinburgh and consisting of 28 members of staff, five of whom are administrative. Five architects head the practice as corporate directors, three of whom have specific duties — John Spencely, chairman; Tom Bostock, the managing director and Neil Gillespie, design director.

The firm works in a variety of construction sectors covering most building types. Each project team is led by a project manager and supported by the computer manager, Tracy Drysdale who co-ordinates all computer activities and is directly responsible to the managing director. This serves to create a consistent approach. All projects involve working within the given constraints of a brief, site, budget and timescale. The practice design policy states

We believe in a simplicity and clarity of approach in order to achieve the client's objectives in the most direct and appropriate manner. As well as ensuring that the project is likely to be affordable and achievable from the outset we believe this approach realizes elegant and refined buildings. We enjoy the consultation and building process itself, materials and construction and as Scottish architects we are acutely aware of our unique location as a generator of ideas.

In terms of design policy our principle concern is with the quality of thinking behind the original conceptual approach and its implementation into a building. All design stages are subject to a review process principally at inception, outline proposals, scheme design, detail design, production information and during construction. This review process is controlled centrally by the directors responsible for design, Neil Gillespie and Bob Steel.

Reiach & Hall Architects were the first practice in Scotland to invest in CAD systems by buying the 3D modelling system — RUCAPS in the 1970s. RUCAPS was successfully used to produce drawings for the Borders General Hospital. In 1990 the firm purchased three Sonata workstations on Silicon Graphic machines, at a total cost of approximately £2000 each. By September 1995, when projects such as the Stirling Centre for Further Education were beginning, it was clear that to allow the office to develop properly and become more efficient it would be necessary to replace the existing drawing system. A decision was taken to upgrade the system again. Reflex was available to replace Sonata, but at a cost of approximately £12 000 each. This time ten Bentley System's MicroStation PowerDraft workstations on RM PCs were installed after consultations with other practices including Richard Rogers & Partners. The aim was to rationalize CAD systems within the practice as well as to have a machine for every workstation, a goal which has since been achieved. Furthermore it was proposed that all technical staff should be able to operate the new drawing system to return direct design control to the designer and remove the necessity of CAD operators.

The intention at the moment is to use the MicroStation PowerDraft for all projects, although some older projects are still using Sonata, constrained by the fact that only a limited number of staff now have

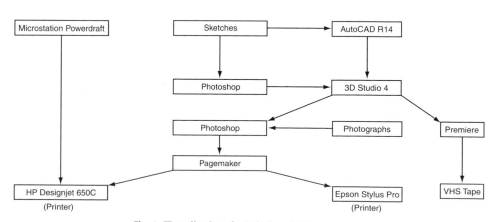

Fig. 1. IT applications for Reiach and Hall

the necessary expertise to use the software. Further purchases of both computer software and hardware are now reactive to needs.

IT applications for text manipulation, databases and specification writing

For administrative computing the practice uses Microsoft Office Professional incorporating Word, Access, PowerPoint and Excel. Both Excel and Access have been used for timesheets, marketing contacts, drawing issue lists and schedules, lists of consultants, clients and suppliers (contacts database), floor area calculations, project information and programming, etc. Microsoft Word is available for use in faxes, reports, minutes and general project administration together with the full NBS format specification system on disk in Word.

Special applications software includes an accounting system; Pagemaker for composing images and drawings for presentation documents; Photoshop for enhancing scanned images, enhancing and photomontaging computer techniques and creating maps and textures for use with 3D Studio; Premiere for editing and composing computer animations, exporting animations to VHS tape and capturing still images from video for presentation and Project Commander. The relationships between the various activities/applications are indicated in Fig. 1.

IT applications for CAD

CAD within the practice uses the three Sonata workstations; the ten MicroStation PowerDraft workstations; and one AutoCAD R13 workstation with an HP Designjet 6500 for plotting.

Originally the Sonata workstations were used principally to generate plans, some sections/elevations and occasionally 3D

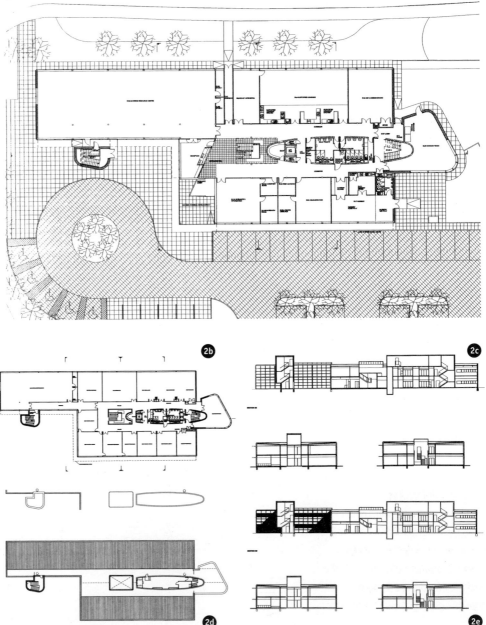

images for presentation purposes. The problem was that users had to have an interest in programming. Sonata has been used extensively to produce drawings for a J. Sainsbury plc superstore project at Craigleith and for the Oban Hospital Project.

New projects are carried out mainly on the MicroStation PowerDraft workstations for 2D drafting purposes. However, AutoCAD and 3D Studio 4 enable importing and exporting information to and from consultants, 3D modelling, computer visual-izations/renderings, animations/fly-throughs. A user group of about four members of staff has been set up to form opinions and advise on graphic standards or to give general guidance.

IT applications for telecommunications and networks

The firm runs Windows NT on a file server. The entire system is networked within the office and facilitates transfer of information by DXF or DWG file types via disk or E-mail. One machine is dedicated to CC mail. Both E-mail installed on one workstation and an Internet link on another workstation are controlled by the computer manager.

An important issue concerns information exchange. To avoid project delays, responsibility has to be defined early on in terms of who produces the initial computer model, who provides upgrades and monitors the model to allow consultants to produce their own information in time.

IT training

Initially, Bentley System's MicroStation PowerDraft training was provided by the system suppliers DATACAD, Hamilton in the west of Scotland. Further training is organized by the computer manager

as required. This may be based upon expertise developed in-house or upon contracted external specialists.

Selected projects

Stirling Centre for Education, Kerse Road, Stirling

A two-storey, 2723 m², educational institute for Falkirk College utilizing the Government's Private Finance Initiative building procurement method. The project was started in September 1995 around the same time as the office began transferring to the MicroStation PowerDraft system. Now the project represents a significant milestone which allowed the firm to learn, iron out diffi-culties and misconceptions while becoming familiar with the computer system as a design tool.

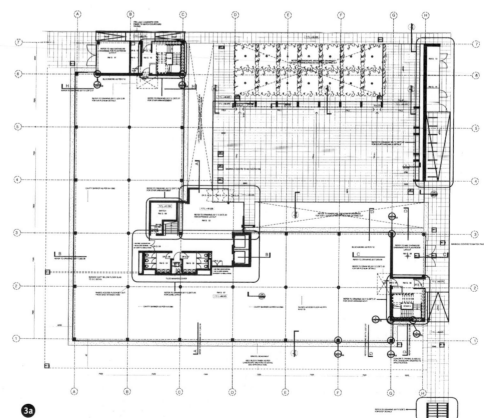

3a

2. (a–e) Initially design drawings (plans, sections and elevations) were done using manual traditional hand-drawn techniques but were redrawn using the new MicroStation PowerDraft system. Files were organized to allow easy subsequent reproduction of the 2D construction information and presentation material. The computer generated drawings (plans, sections and elevations) were notable for their clarity and consistency of information allowing feedback from the project which has also become a model for all other projects. The drawings were used for planning permission, building warrant, builders' working details and for exhibition purposes.

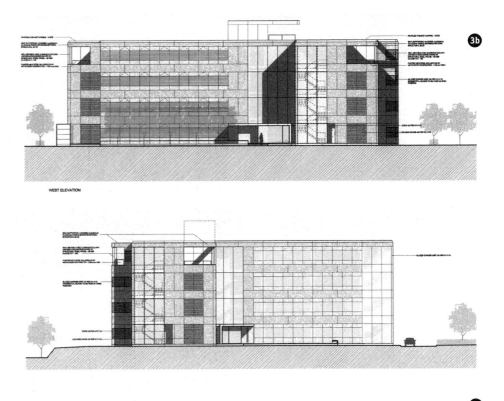

An office development for New Edinburgh Ltd, Site E2, Edinburgh Park

This project was started in mid 1997, by which time the office had been using the MicroStation PowerDraft system for over one and half years. By this time, staff had acquired the essential experience and showed a new level of competence and remarkable confidence with using the technology. The simple, open-plan, 3150 m² building with its high level of repetition over four floors enclosed by a taut, elegant 'skin' allowed the office the opportunity to use the computer system at all levels from the start of the project — for initial design presentation material including 3D visuals and animations, for 2D working drawings and for project management.

3. (a–c) Before beginning any drawing work, the project team set out how the computer is to be used in order to control the tendency of information to proliferate and duplicate. A strict hierarchical file structure is created and templates for various aspects established including provision of a resource of details to save drafting time. The plans, sections and elevations and details indicate the nature of the computer drawings

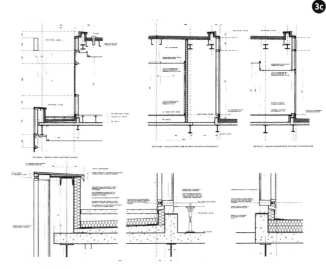

The project allowed the office to invest in visualization and video editing software and hardware for the purpose of preparing presentation videos of the exterior and interior of the building for the client, letting agents and planning authorities

RHWL Ltd, Architects, Planners and Designers | 22

Background

RHWL was formed 32 years ago under the founding partners Renton Howard Wood and Levin. RHWL has one main office on four floors located in London on Endell Street and a satellite office is situated across the road. The practice comprises 120 staff — 70% architects and 30% administration and technical support. The practice has a number of partners who each lead a group making up a specialist area, although in many situations there may be some cross-over in the nature of work and operation. The main groups include

- arts
- leisure
- commercial
- mixed development
- central resources group.

IT matters fall within the remit of the central resources group and have three main elements.

(a) The IT group comprises the IT manager (Nick Dunn) and an architectural technician who acts as a part-time assistant. Nick Dunn has a background of 15 years in IT, 10 of which are with CAD systems within the practice.

(b) The technical department is responsible for technical aspects within the practice including specifications.

(c) The graphics group is responsible for 3D perspectives most of which are still hand drawn and increasingly are being based on CAD outlines.

RHWL's strategy on IT is based on the belief of 'keeping it simple'.

IT applications for text manipulation, databases and specification writing

RHWL uses Microsoft Office Word for word processing including aspects such as naming conventions, standard document templates (e.g. refer to typical example of query sheets; minutes of meeting template; architect's instruction) and standard RHWL specification with reference to NBS. Excel is used for area schedules programming/programmes, PowerPoint for presentation material and Access for accounting purposes by the accounts department. Tetra software is used for some accounting activities (fees/losses, etc.) and has bespoke connection to Access. Timesheets are handwritten before being passed on to the accounts department.

The database software, File Maker Pro, is used for door/room schedules, making minutes/notes of meetings, checklists, drawing issue lists instead of sophisticated document management systems

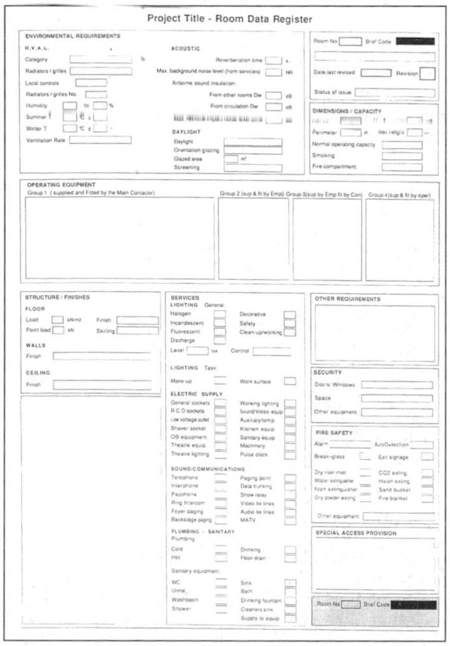

Fig. 1. Room data register

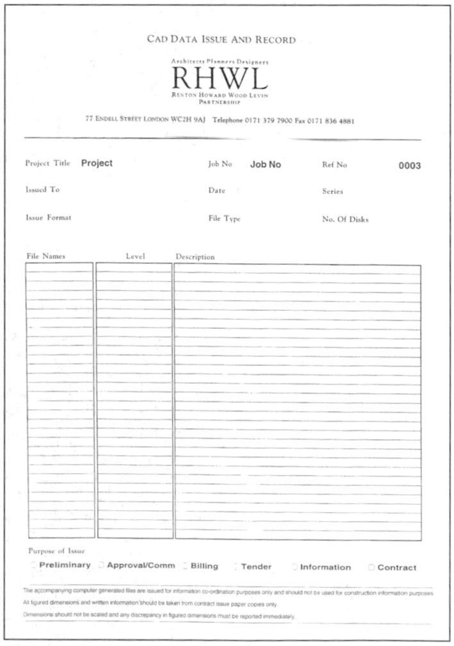

Fig. 2. CAD data issue and record

(see the typical example: Project Title — Room Data Register (Fig. 1); CAD Data Issue and Record (Fig.2)). QuarkXpress/Photoshop are used for presentation material, brochures and reports incorporating images.

IT applications for CAD

RHWL use the full version of MicroStation based on 35 licences. The practice has between 60–70 machines (95% PowerMacs) purchased within the last five years including a more recent purchase of the latest model with lots of memory for a particular project. The accounts department has five Pentium PCs. Other equipment includes one A3 scanner; three Hewlett Packard Design Jet plotters (1 colour and 2 black and white); three A3 laser printers, one ink-jet printer; one A4 DAfoundal ruler 10 x 8; several A4 laser printers and three NT servers. Other software packages include Form Z for the arts team and 3D MicroStation Modeller-Core.

Figure 3 shows the nature of the RHWL process which starts with the concept followed by design sketches and design using Form Z modelling and on-going design work ending up as paper copies or electronic (E-mail) copies for communication with other members of the project team. On-going design consists of drawing board work mostly by the graphics group or bespoke MDL (MicroStation

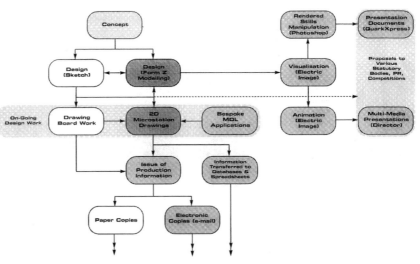

Fig. 3. The design process from concept to presentation

Development Language) applications all of which lead to 2D MicroStation drawings used for general plan/layouts developments, including sections and elevations, to produce production information or information for transfer to databases and spreadsheets.

Form Z modelling leads to visualization (rendered stills manipulation using Photoshop or Animation using electric imaging) to produce proposals for various statutory bodies, PR, and competitions. Visualization/animation work although increasingly done in-house is still often outsourced to John Bell at the University of Westminster or POD Communications for Quick-Time VR/videos or 3D output as requested by the client to suit specific project needs.

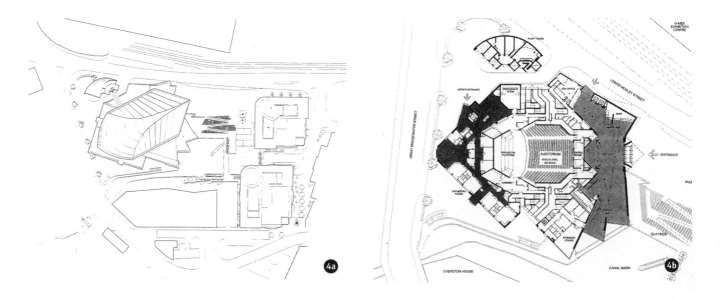

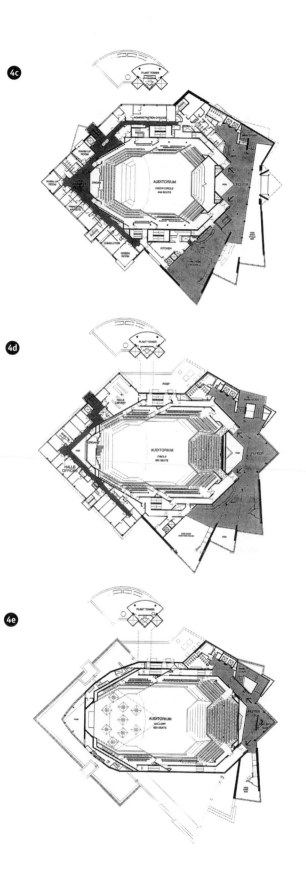

IT applications for telecommunications and networks

RHWL network consists of ethernet netware (thin wire upgrade structure cabling) with Appletalk; dial-manual link and 2 megabyte lease line link to the satellite office. The Internet (modem) link allows integration with a fax (modem mail and fax servers). Internal Microsoft mail with fax gateways is used.

IT training

RHWL have a policy of employing people with suitable computer skills. Originally staff were provided with training on the in-house training scheme. Subsequently, additional training has been provided for the practice by Bentley Systems Business Unit using six day sessions on consecutive Saturdays. Rothwell Group (Micro vendors for Apple Macs) have provided support following purchase and delivery of the equipment from them.

Other activities include those of the accounts group which meets once a month and acts as a focus group or talking shop.

Selected projects

An Indoor Ski-Centre, Taiwan

In this project early outputs included a video requested and paid for by the client for an Open Day Celebration to promote the project.

Bridgewater 2400-seat Concert Hall Auditorium, Manchester

The client for this project was Manchester City Council. The landmark building is part of Manchester's Great Bridgewater initiative, regenerating this central district of the city. The stone clad auditorium is enveloped by elegant foyers which oversail the adjacent canal basin. A dramatic building which symbolizes the city's commitment to the arts

This is a high quality space... Manchester has gained a striking new urban element... RHWL's achievement has been to create a building that meets stringent technical demands, and one that sits comfortably in its context.[1]

4. Site plan (a) and Level 1 stalls (b) showing the entrance from Lower Mosley Street. Levels 2–4 1st tiers–3rd tiers (c–e) show the building as composed from a series of sculptured tiers, combined with a classic shoe box hall. This enables the audience to enjoy a close aural and visual contact with performance, giving a sense of involvement with the music making. CAD tools aided drawing of sculptured layouts.

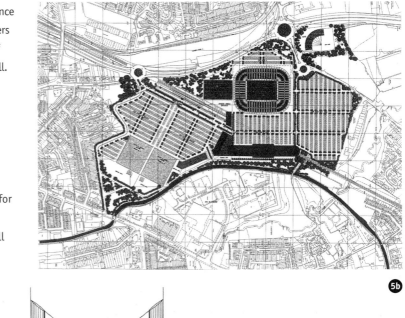

National Arena Super Stadium

This stadium with a maximum capacity of 40 000 seats for football matches rising to 45 000 for a pop concert is situated on the Foleshill site, Coventry. The stadium will be the first of its kind in the UK incorporating both a retractable pitch and a retractable roof, making it an all-year round venue for football, other sports, concerts, exhibitions and conferences. The stadium houses adjustable advertising screens which are able to block out part/all of the upper tier. At a cost of £42 million to construct and £18 million to fit out the Arena is designed to provide parking for 6000 cars and 300 coaches. It will be served by rail with a dedicated station on the Coventry to Nuneaton Line.

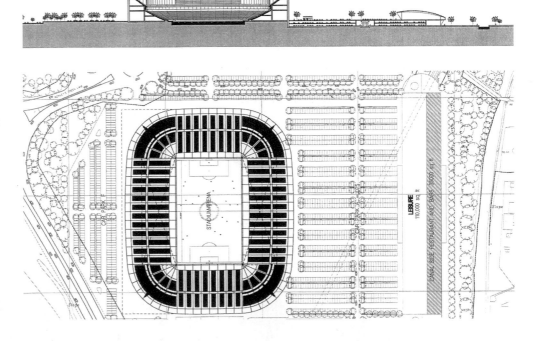

5. 2D CAD layouts
   (a) masterplan layout
   (b) ground plan and section

Reference

1. Stungi N. Popular classic: RHWL's Bridgewater Hall — Manchester's new venue for concerts; Architects: Renton Howard Wood Levin Partnership. *RIBA Journal*, 10, profile supplement, Oct., 1996, 3–29. RHWL profile by Ken Powell, 11–13.

| **Richard Rogers Partnership,** Architects | **23** |

Background

Richard Rogers founded the practice with John Young, Marco Goldschmied and Michael Davies in 1977.[1-3] The senior directors have been working together for over twenty-five years and provide the conceptual basis for the projects. Richard Rogers received the Royal Gold Medal for Architecture in 1985, a knighthood for services to architecture in the Queen's 1991 Birthday Honours list and a peerage in 1996.[4, 5]

The Richard Rogers Partnership now has 100 staff including 10 directors, 6 associate directors, project architects, administration and a computer support team. The Richard Rogers Partnership is a limited company, wholly owned, and is run on a profit-share basis with 20% of the architects' profits allocated to charities of their individual choice.

The practice has built a wide range of projects from low cost industrial units to prestige headquarters; highly technical laboratories to sensitive landscape proposals; cultural centres to speculative office developments; airport planning to restoration of historic monuments. As much as 50% of the commissions are for projects that are from abroad — Germany, France, the Far East and the USA — and almost a third of the work is for commercial developers.

Recent commissions have included: the design of two law court facilities, the European Court of Human Rights in Strasbourg and the Law Courts in Bordeaux; the conversion and restoration of

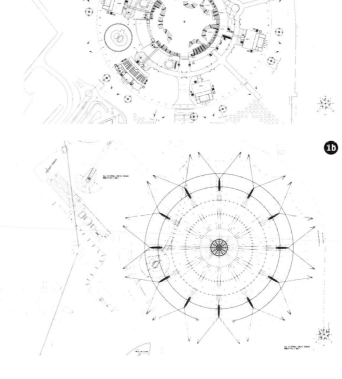

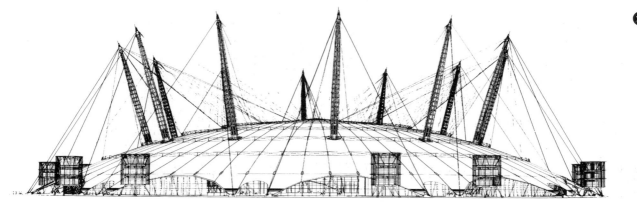

2a

Billingsgate fish market in the City of London; the new headquarters for Channel 4 Television; the Euro-pier Terminal at Heathrow and Heathrow's Fifth Terminal; masterplanning for Potsdamer Platz, Berlin; City of London headquarters for Daiwa and Lloyd's Register of Shipping.[6] The practice has responded to growing concerns for the environment and for enhancing sustainability by developing research into energy efficient buildings for current schemes in Majorca, Spain, Berlin and London.

Throughout the history of the practice the theme of masterplanning — creating public space and enlivening the city — has played a significant role in shaping designs. This theme has been very successful and is evident in the Pompidou Centre Project, a project won in competition, and it was proposed that over half the site area be given over to the creation of a public piazza.

The practice's reputation for thorough analysis and innovative problem-solving has been gained through often exhaustive consideration of each project from first principles. Lloyd's of London required a building which could be capable of leading a then expanding insurance market into the 21st century. A range of options were presented and rigorously debated with the client at regular meetings, until the final solution to specific demands of the brief was agreed. A strategy was adopted which allowed the market place to expand and contract according to the unpre-

dictable market forces, the same space accommodating eight times as many people when performing as a market as when converted to normal office space.

All four partners work on the design evolution with a director and project architect. There are weekly design reviews at which entire design teams present the work to the partners for discussion and criticism, a process which continues throughout the design and ensures the quality of the final building.

Lennart Grut is the director responsible for IT development and Peter Davenport has administrative responsibilities with the Richard Rogers Partnership. Ian Martin Associates Consultancy provide the strategy and operational guidance with Simon Williams-Gunn, an employee of Ian Martin Associates, as CAD manager but resident at the Richard Rogers Partnership premises. Mr S. Williams-Gunn was formerly with Allies & Morrison Architects and holds a master's degree in conceptual computing. Erica Reeve is involved in providing support for the management database (on in-house based 4 Dimension software) for accounting purposes.

The practice policy is not to employ CAD technicians, aiming to develop conceptual entities as tools defined by light, space and form rather than a mere concentration on corporate identity. The Richard Rogers Partnership had an in-house CAD group which developed its own software but this had proved expensive to resource hence the adoption of the management arrangement involving Ian Martin Associates Consultancy.

Currently the Richard Rogers Partnership are reappraising the position of IT within the practice driven by a number of factors notably

- the size and complexity of projects involving multi-disciplinary teams and in which communication and information flow are important
- the size of client organizations, for example Mitsubishi, Japan whose IT capabilities are huge and in which IT platforms for inter-operability and better exchange of electronic data between members of the construction team are critical.[7]

The IT strategy involves a continual programme of upgrading hardware/software as well as raising the technical competence level of each of the teams. The need is to move as quickly as possible to NT servers, in what they perceive to be an increasingly 'NT world', and construction of WANs especially when they may have to employ say 20 site architects in some multi-disciplinary major project.

IT applications for text manipulation and databases

The Richard Rogers Partnership uses Microsoft Office based on need: Word on the Mac for word processing, Excel used for timesheets, Microsoft Mail, etc. Specification writing is outsourced.

Other software includes QuarkXpress/Photoshop for brochures and report writing; FileMaker Pro used for automation of drawing issues and timesheets; testing Cumulus for doing PR information; 'Now-to-date' software is used for directors' diaries with both private and public entries; considering 'paperless office' by Paper Port Strobe for running jobs and to be plugged into the Internet.

IT applications for CAD

The Richard Rogers Partnership has 40 licences to run MicroStation (complete suite of MasterPiece, Triforma, etc.) from Bentley Systems and 40 CSP. The hardware comprises AppleMacs: 2CII PowerMacs (7600, 7300, etc.), Silicon Graphics machine, and one 266 Pentium PC. They also have two copies of AutoCAD. Recently a decision was made that any new machines are to be PCs. They would hope to review this when Mac's Rhapsody with the 'plug and play' characteristics becomes available.

Other software includes Form Z for 3D design, Lightscape and StrataStudio Pro.

The firm also outsource presentation/visualization (video, etc.) material production to market-leaders such as Hayes Davidson. The Richard Rogers Partnership process involves

(a) conceptual design based on models and sketches often by hand from the founding partners who appear not to use computers extensively

(b) development of presentation material and therefore building of a 3D model

(c) development of schemes through planning activities mostly using 2D drafting and a few with 3D modelling

(d) development of technical details mostly using 2D drafting.

IT applications for telecommunications and network

The Richard Rogers Partnership have five ISDN lines as links to the outside; 4Sight — five routers; Fritz — client link to Frankfurt and Sarkpoint — link to a reprographics company for specialist printing. There are seven servers using Apple share and one Windows NT but the aim is to migrate totally to NT servers eventually. The fax is 'Freefax' but there is no electronic fax. The web site location is *http://www.richardrogers.co.uk.*

IT training

Training is provided on a need basis with in some cases a 'lot of hand-holding'. Ian Martin Associates provide any training and support as required otherwise the policy is to employ computer literate staff.

Issues of concern to the firm and to the CAD manager include

- data ownership; accuracy (e.g. CAD accuracy or driven by the need for accurate dimensioning of drawings)

- responsibility for export to other members of the project team (definition is required in terms of cost and time)

- role demarcation and clarification of relationships between consultants

- barriers to effective use of IT include culture based on old methods of working and expectations, e.g. satisfying contractors' needs for drawings and building details. The need is for change of perception such that the 'building' is an object from which drawings are produced. This means when an architect sets up a wall, a wall object will be created, with physical characteristics such as length and thickness and further attributes for material layers within the wall. This would include, for example, U-value, sound absorption coefficient and its colour. All of this object information would then be stored with the wall object segment, and would be made readily available to the whole of the construction team

- doubling-up of electronic and paper copies.

Selected projects

Four projects have been selected as representing significant milestones in connection with IT developments within the practice: the £750 million Millennium Dome in Greenwich, the Lloyds Register of Shipping headquarters in Surrey,[8] the £600–800 million headquarters for Nippon Television in Tokyo and the £175 million Madrid Airport Project in Spain.

The Millennium Dome

This project with Buro Happold (engineers) and McAlpine and Laing (Project Managers), required communication links between the site office and the consultants. This had to overcome cross-platform file transfers and had to ensure all parties were involved and were working electronically together. These needs encouraged the practice to install the software Model Server Publisher by Bentley Systems. For Richard Rogers Partnership the other projects reflect the enormous changes taking place within the practice in relation to the use of computer software and hardware and they also signify the increasing relevance of IT tools and enabling technologies, especially in situations where big corporate clients are involved. For most projects, design, creativity, technology and appropriate management are the essential ingredients of success.

1. (a–c) 2D CAD layouts (plans, elevations etc.) depict the world's largest supported dome at 320 m in diameter and 50 m high at its centre. The dome is described as being able to fit 2 Wembley Stadiums, 13 Albert Halls, 3300 London buses, etc. The dome covers exhibits as 'a focus and identity for the New Millennium Experience (NME) — its roof will be white, its masts 'Van Gogh cornfield yellow', a feat of today's technology 'that hardly touches the ground'

2. (a–c) 3D CAD images complement physical models

Barajas International Airport, Madrid, Spain

The £8.4 million New Area Terminal (NAT) of Barajas International Airport was won in association with a local architectural firm, Studio Lamera (airport engineering consultants TBV Partnership and engineers Intec), in a competition of 20 practices. The 20 000 m² NAT is the largest in Spain.

3. 2D CAD section through the New Area Terminal of Barajas International Airport denotes the type of drawings being produced for the project

Headquarters for Nippon Television, Tokyo, Japan

Richard Rogers Partnership with Mitsubishi Estate Co. won the competition to design a 115 000 m² building in Tokyo for Nippon Television headquarters. The new headquarters will be approximately 185 m high and situated to the east of Shinbashi JR station in the south-east Tokyo district of Shiodome, which overlooks Tokyo Bay. The building will include television studios in the lower floors and naturally lit offices above. In the heart of the offices will be glazed atriums for internal circulation, which will also allow views across the Tokyo skyline. As part of the complex there will also be an art gallery, shops, restaurants and public auditorium adjacent to an external public space. The building will be constructed from steel and concrete with twin glazed façades for solar power and comfort control within the offices. Television satellite dishes, mounted on the roof, will allow the company to operate 24 hours a day. The building has been designed to be linked to the neighbouring hotel that is to be designed by the Kajima Corporation and built at about the same time. Adjacent to the whole site is a mixed office redevelopment area designed by Kevin Roche and the new Denstu HQ designed by Jean Nouvel.

4. Typical 3D CAD images being produced

References

1. Chevin D. Jolly Rogers: interview with Sir Richard Rogers. *Buildings*, **261**, No. 7944 (23), 7 June, 1996, 42–43.

2. Appleyard B. *Richard Rogers: a biography*. London, 1986.

3. Sudjic D. *The architecture of Richard Rogers*. 4th Estate. London, 1994.

4. Lewis J. Rogers gets life peerage. *Building Design*. No. 1275, 23 August, 1996, 1.

5. Futagawa Y. Special Issue: Richard Rogers. *GA Document Extra*, No. 2, 1995, 10–168.

6. Burdett R. (ed.) *Richard Rogers Partnership: works and projects*. Monacelli, New York, 1996.

7. Sudjic D. *Nine projects for K-One Corporation and Mitsubishi/Richard Rogers Partnership*. Wordsearch, 1991.

8. Melvin J. Where the skies meet the street: Lloyds Register of Shipping, Headquarters. *Building Design*, No. 1242, 10 Nov., 1995, 8.

SOM, Architects

Background

SOM (Skidmore, Owings and Merril) was founded in Chicago[1] and has consistently been in the top ten American architectural practices. The firm has four major offices worldwide with a total staff of 800: Chicago SOM HQ with approximately 350 staff (architects, planners, interior designers, structural, civil and mechanical engineers); New York; San Francisco with approximately 200 staff; London with around 50 staff (primarily architects, planners and interior designers) and branch offices in Los Angeles and Washington DC. These offices share expertise while being able to compete for work.

SOM's mission statement[2] identifies the following goals of the firm:

- to achieve excellence in designs that serve clients and communities

- to understand each client's specific aspirations and needs, and to provide service of value

- to lead the profession by developing forward-looking and innovative approaches to design, technology and management

- to practice in multi-disciplinary teams that can aspire to achieve greater quality and success than individuals working alone

- to provide a stimulating and rewarding work environment that encourages the professional growth and development of individuals throughout the firm

- to foster the vitality and renewal of the firm.

While the US practice is a partnership with limited liability, SOM London is incor-

porated. The SOM London office was set up in 1986 on the back of the job to masterplan Canary Wharf and work on the Broadgate development. Important SOM work in the UK includes the Heinz building, listed Grade II, and the Boots headquarters in Nottingham. SOM London director is Roger Kallman.

SOM is a multi-disciplinary company with areas of expertise in architecture, structural (historical connection), mechanical and electrical engineering, planning and interior design. The range of work involves large scale commercial developments such as Canary

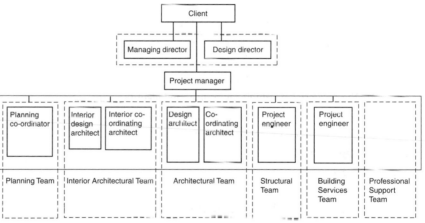

Fig. 1. The project team

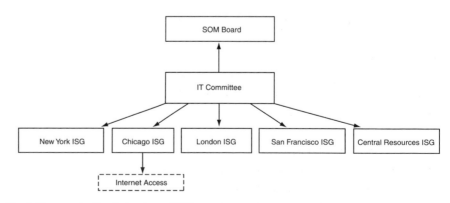

Fig. 2. The organizational structure of SOM

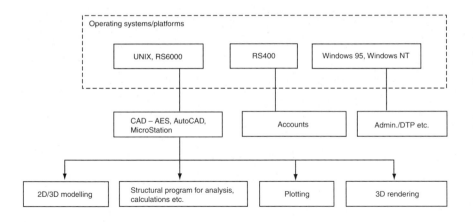

Fig. 3. IT applications at SOM

negotiating the contract, assigning resources and monitoring the programme. The day-to-day operations are the responsibility of the project manager.

IT tools are important for a firm such as SOM. The need is for the various offices to respond to unique local requirements and conditions yet follow a similar IT development path to ensure 'firmwide compatibility'. New IT equipment (hardware and software) has an annual budget of approximately $1.4 million, that is $2000

Wharf (1985–1991) and Broadgate Developments (1987–1990); a tower is classic SOM.[3] Project leaders are responsible for projects and each partner is responsible for each project. The firm works across diverse building types, such as airport projects, offices, and financial institution buildings in order to cover fluctuations in market sectors. The main sources of work are repeat business, new clients through marketing and limited competitions.

Project leaders are responsible for projects and each partner is responsible for each project. For each new project, a project team is established comprising personnel drawn from various disciplines so that individual skills and abilities are matched to the aims and objectives of the client. The working relationships are indicated in Fig. 1. A senior designer, technical co-ordinator and project manager form the nucleus of every project, under the direction of a design director and managing director, both of whom are partners. The design director has overall responsibility for the design and will work with the senior designer identifying, refining and translating concepts and ideas into a physical form. The managing director has responsibility for project management, establishing fees,

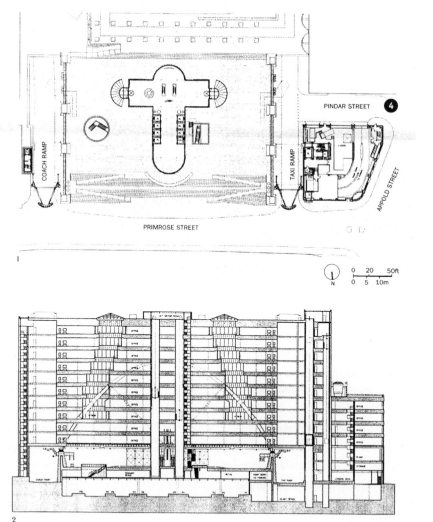

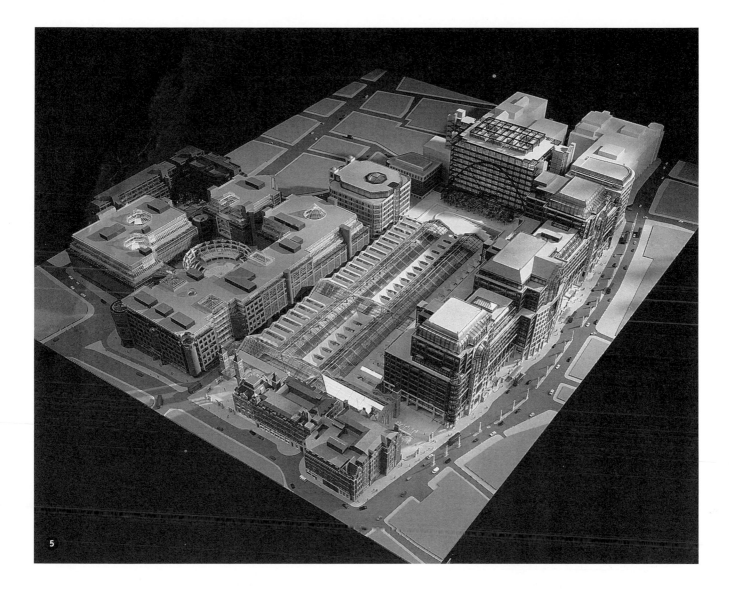

5

per person (compare with the industry average expenditure of $3000 per person). The IT basis of SOM is the corporate UNIX platform, global E-mail and total compatibility between offices. The IT organization structure is shown in Fig. 2. Each office has an information services group reporting to an IT committee who in turn report to the SOM Board.

In the US the project process involves (a) architectural activities using IBM's AES (Architecture Engineering Series) system developed with the assistance of SOM[4] and (b) technical activities using AutoCAD for 2D drawings (as it is easy to find experienced users at

this work stage). In the UK the project process is a single one with DXF transfers to either AutoCAD or Microstation. The method of work involves physical models, freehand concepts development (later scanned for records purposes) and use of computers.

Charles Davis, information systems manager in London formerly with Scott Brownrigg & Turner for two and half years has been with SOM for a year implementing IT integration within the office. He explains 'SOM has always been at the forefront of introducing Information Technology into its design process as well as in its business systems'.

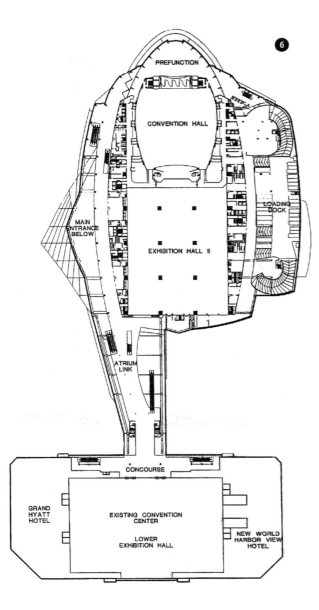

letters, faxes, etc. Excel is also used as accounts' software which covers FMS on AS400 accessed by Netsoft client access for PC and Showcase Vista. Timesheets are done manually and there are plans for direct electronic input.

Quark Xpress, US/Photoshop and Pagemaker are used for marketing purposes while specification writing is carried out by specialists in the Chicago office using TBA (Custom system). More SOM firmware consists of Windows 95/NT operating system (replacing UNIX); Winzip US FTP and Eudora E-mail.

IT applications for CAD

IT development and the need to track, control and reuse data efficiently are important for SOM. Since the mid 1970s SOM has invested heavily in IT with the hold copyright. AE&S is now used widely in the US (in offices such as Richard Meier') and as far afield as Australia, India, Japan, etc. In 1983 SOM purchased an Intergraph CAD system for their Houston office, although the firm's offices were using AE&S, SOM's home-grown CAD software. Autodesk 3D Studio, Intergraph Modelview and Form Z are used extensively during CAD applications and an iterative design process (Fig. 3).

AE&S with its four programs (Model, Plot, Sheet and Render) is embedded in the SOM design process offering flexibility and suitable gearing in particular because it is a multi-user system, allowing up to 100 persons to work on a project at any one time, and also gives the capability of working in both 2D, 3D and colour.

IT applications for telecommunications and networks

For a multi-disciplinary firm like SOM communication links are essential due to the size of the firm and diverse range of clients and their projects. This is facilitated by the integration of all platforms (UNIX, AS400, PC and Mac) on TCPIP involving many features — a central E-mail system (using Eudora), firm website, FTP site/Bulletin Board, office intranets, international remote access, frame relay WAN, all networks either tokenring or ethernet and a central marketing database.

IT applications for text manipulation, databases and specification writing

All administration and accounting software and operating systems have been standardized on PCs throughout SOM's offices. Microsoft Office Professional (Word, Excel, PowerPoint and Access) is the SOM administration standard software and one used for templates for all

SOM website is at *SOM@compserve.com* which carries the message: 'We deal with the oldest forms of man's concern: shelter, one even more important, his need for beauty and personal expression'.[6]

IT training

Training is an important issue for the firm because of the need to understand large amounts of hardware/software and various systems in order to utilize the information systems to exploit their full potential. The aims are to increase staff knowledge — from basic computer skills (Microsoft Office and E-mail) to CAD (AES, AutoCAD and Intergaph) — and to do so on a continuous basis and at a certain level in order to reap the full rewards of IT investment.

SOM policy on IT training is dictated by the budget. PC and administration training is provided as a three/four day session every quarter using external consultants on the office premises. DTP is provided using both in-house and external trainers while AE&S and CAD training is provided in-house — twice a year as a one week session.

Charles Davis sees the main problems as being

- low staff computer literacy and therefore the need for training by the industry
- the need for continuous funding/investment
- the need for more project co-ordination including putting the data in the right place at the right time.

Selected projects

Three projects have been selected to reflect the crucial role of IT and other technological developments for the firm. These large projects, or rather massive undertakings, are characterized by massive scale, staggering complexity, very tight programmes requiring creative solutions which include the need for consistency, uniform quality and communication across many design teams and construction participants often from many different countries (e.g. US, Hong Kong, Australia, etc.) In most of these projects the Internet is important to get information from the designers to the builders. Procedures are adopted to standardize models, distribute and track information. The projects are Utopia Pavilion Lisbon Expo 1998 in Portugal, Exchange House in London and the Hong Kong Convention Centre Extension project.

In the Lisbon Expo project IT has been helpful in the master-planning project. In the Exchange House and the Hong Kong Convention Centre Extension, current technology was used to achieve sophisticated functional designs of landmark buildings.

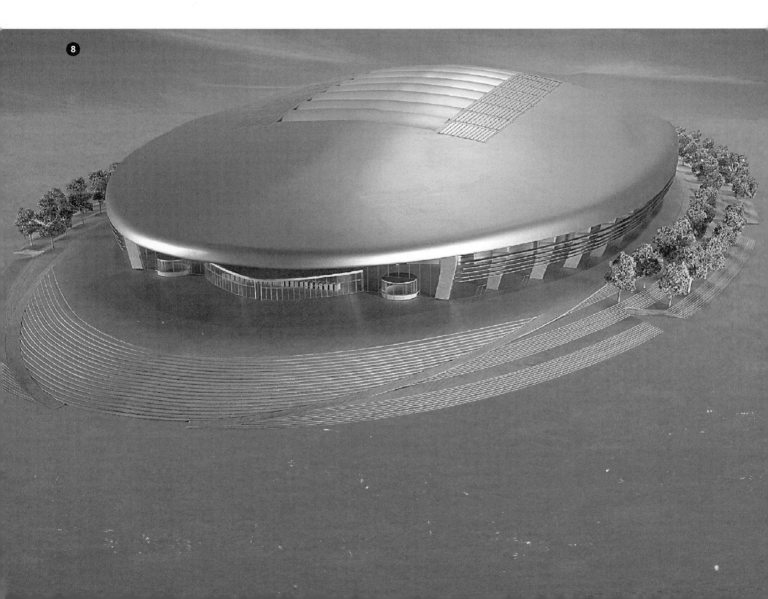

8

Exchange House, London

Broadgate multi-use development was the result of a design competition in 1987 and comprises ten buildings of which Exchange House is one. The ten buildings are designed in a variety of styles to relate to the City context and are grouped to form public squares with the terraces and the landscaping of the Exchange Square providing the focal point of the development. The ten-storey office block supported on an expressed structural frame that spans the Liverpool Street Station tracks is the Exchange House.

4. 2D layouts (plaza-plan level and east-west section)
5. 3D CAD aerial view of the Broadgate development with Exchange House at the top

Hong Kong Convention Centre, Hong Kong

Designed by SOM in association with the local architecture and engineering firm, Won & Ouyang Ltd, the 148 640 m² extension to Asia's first integrated convention centre (opened in 1988) is full of symbolism — the sculptural roof, the irregular footprint shapes, maximum transparent façades, etc. The civic land mark, built on a 6.5 hectare man-made island with more than 90 000 m³ of sand-fill, is dominated by a 40 000 m² sculpted aluminium roof symbolizing a sea bird in flight —'a gliding manta-ray'.

6. 2D CAD layout shows the multiple-use spaces, organized into a tight footprint prescribing vertical emphasis with 3 exhibition halls stacked on top of each other and a 4500 seat convention hall at the summit. The convention hall and the top-level exhibition hall occupy an 80 m wide-span space under a sweeping 14 m high curved ceiling. Also with the multi-storey glazed circulation around the periphery, a glass-and-steel curtain wall façade emphasizes transparency and exploits breathtaking views of Victoria Harbour

The complex, which includes 5 exhibition halls, 2 convention halls, 2 theatres, 52 meeting rooms and 7 restaurants, housed the 1997 handover ceremony of Hong Kong to China. The 3D CAD illustrations of the roof whose visualization was facilitated by using 3D engineering software and the curved layered elevational treatment help enforce a sense of the horizontal thereby striking a contrast with the largely vertical city behind.

Utopia Pavilion, Expo 1998, Lisbon, Portugal

The 30 000 m² multi-purpose pavilion project was another competition winning design. The design, which includes a six-lane, 200 m Olympic track seating 17 500 spectators, is capable of accommodating a full range of indoor sports and events such as concerts, conferences and exhibitions. The low rake of the spectator seating within the arena enhances the spectator's views, making the facility quickly adaptable for musical and theatrical events.

7. (a–e) 2D CAD layouts (location, site and floor plans, elevations, sections and structural details). The structural illustrations confirm the ribbed roof structure of laminated timber beams manufactured from softwood, the distinctive massing and the resulting aerodynamic form. The building form not only relates to the history of the site which used to be a landing site for seaplanes but also creates a favourable wind flow allowing the building to have natural ventilation for energy efficiency.

The pavilion, whose shape draws inspiration from the lines of the carabelas, the ocean-going sailing ships of Vasco da Gama's day, was designed to be the centre-piece of a 200 hectare urban riverfront regeneration.

8. 3D CAD drawings.

References

1. Gretes, F. C. *Skidmore Owings & Merrill 1936-83*. Monticelli: Vance Bibliographies, 1984.

2. Curtis, C. and Gates, C. *The International Directory of Architecture and Design*, **4**, DID Publishers, London, 1994, 237.

3. Skidmore Owings & Merrill. *Architecture of Skidmore Owings & Merrill 1984 -94*. The Master Architect Series, The Images Publishing Group Pty Ltd, Mulgrave, Australia, 2nd Edition, 1995, 18.

4. *Ibid.*, 138.

5. Giorgio, F., Il CAD di Skidmore Owings & Merrill (CAD by Skidmore Owings & Merrill). *Arca.* No. 39, June, 1990, 62-69.

6. Bush-Brown, A. *Skidmore Owings & Merrill Architecture and Urbanism 1973–83*. London, Thames & Hudson, 1984.

7. Owings, N. A. *The Spaces in Between: An Architect's Journey*. Boston, 1973.

Part three Conclusions and future developments

Introduction

The survey into case study organizations sought to gain an insight into the working of the construction

design practice. Questions aimed to discover trends and main issues. This involved looking at

- *Background*: i.e. a brief history of the practice covering the organization structure and how it relates to the office administration, to projects and CAD management. The main question is how is the practice organized and resourced to exploit IT? What are the policies on IT? Are there any IT standards?

- *IT applications for text manipulation, databases, specification writing*: i.e. Microsoft Word for Windows for text manipulation, letter and report-writing; QuarkXpress for desktop publishing work. In addition this means examining other software used, e.g. Sage for accounting; Project Command for timesheets; No Teamate for spreadsheets; or Specification Manager for specification writing.

  The schedules produced covered
  - accommodation
  - floor area analysis
  - room briefing data
  - services
  - room finishes
  - fixtures and fittings
  - furniture
  - equipment
  - sign posting.

- *IT applications for CAD*: e.g. AutoCAD and MicroStation for all aspects of work — for instance all working drawings, 3D modelling, animations and presentation plotting; Intergraph's

Modelview for specialized rendering software. The main emphasis is to show that CAD is more than AutoCAD but relates to the description of the practice's complex process as an idea is developed from concepts to sketches, to 2D, to 3D production information and/or rendering and video for client presentation. Many technologies such as smart sketching and object-oriented drafting are available for production of drawings without meticulously placing each entity on a drawing in much the same way pencils and paper have been used for centuries. In smart sketching concepts are laid down without ensuring that lines are perpendicular or figuring out exact dimensions, whereas in object-oriented drafting the object may alter itself based on changes, e.g. a building truss that automatically stretches and adds more cross supports as the span is increased.

- *IT applications for telecommunications and networks*: i.e. hardware — types and number of machines (e.g. Apple PowerMacs, Pentium PCs, etc.) including type of network and links with other offices or consultants as members of the project/design team; and a description of software — including network software type and other considerations plus future plans. Of importance is the nature of integration with the client base and other professionals.

- *IT training*: i.e. the nature of training, help desk and trouble shooting provisions. For example whether this is via in-house using experienced users, vender support or consultancy.

Computer hardware and software represent a big investment to the firms and to maximize return on the investment requires training and education.

- Discovering what are *the main IT issues* of concern to the practice? (For example rate of change of technology, incompatible data formats, training, legacy, etc.)

IT training is a necessity for the firms. Although project managers do not need to master every software application used by project teams, they should understand at least the following:
  - the organization of the network file system, including hierarchy and file-naming conventions
  - the organization of CAD documents, including layers, components and reference files
  - the scope of the firm's digital knowledge base and availability of on-line resources
  - the pace at which digital information is developed and reproduced on a typical job
  - the logistical issues and liabilities of sharing digital data with clients and third parties
  - the availability of talent within the firm for a given task.

At a minimum, a project manager should know how to
  - locate a document in the computer file system
  - open a document, using the application used to create it
  - perform basic edits (margin notes or redmarks)
  - print or plot a document or drawing
  - comfortably use internal and external E-mail, messaging and on-line services.

- *Selected projects*: Important reference material on IT applications by the practice seeks to show innovativeness and highlights how useful IT has been. The projects are described by details — location/site plans, floor plans, elevation/section, details and computer Images. These are not meant to be comprehensive but to provide examples of how the practice use IT tools in day-to-day operation and from project to project. Also when looking at projects computers should not take all the credit for the design.

Conclusions and future developments

The background information of the case study organizations shows significant differences based upon historical experiences. Reasons for the involvement of architectural and engineering firms in using computer tools were varied, reflecting the separation between (*a*) word processing, spreadsheets, databases, (*b*) financial systems, and (*c*) CAD. These reasons were based on the following aspects.

- Project requirements, e.g. complexity (especially when projects are characterized by the absence of traditional building blocks of beams and columns (shells, drawers, etc.)), examination of options, time and resources constraints with outsourcing as an accepted way for the organization to focus on core competencies, obtain the benefit of specialist skills, keep abreast of market developments and reduce costs. Access to up-to-date product information is also crucial.

- Client requirements and expectations, e.g. a firm of a certain size is expected to have advanced CAD capabilities, visualization and animation tools, project management and scheduling abilities and opportunities to explore alternative design solutions.

- Pressures to maintain a competitive business advantage and the need to find a route to 'world class performance' in order to maintain active superior levels of customer service, e.g. priority is given to obtaining the right person for the job rather than having computer tools which are easier to obtain, as well as IT skills, IT best practices, strategies and policy, e.g. a retirement programme for obsolete IT equipment. Communicating electronically (E-communication) has the potential to enhance competitiveness to enable cultural change and to help organizations capture and use the sum of the knowledge acquired by staff, thereby fuelling innovation and the collection of ideas translatable into new products, services and business processes. E-communication encourages collaboration and dispenses with, for example, CAD standardization — designs from one platform can be translated to another and the entire assembly remains intact with 100% success.

- Office requirements.

In many practices responsibility was delegated to less than enthusiastic senior members. This is consistent with the survey commissioned by the Institute of Directors and by Oracle which found that UK directors of companies are not using PCs, and not only are they failing to grasp the potential of IT, they are actually delegating IT strategy to board members with little or no direct experience of computers. The survey found that only 64% of senior business people in the UK use computers in the office compared with 84% in Germany and 100% in both the USA and Singapore. The only application widely used by UK directors is E-mail and many participants in the survey commented on their lack of knowledge about, or ability to use, a computer.

IT applications for text manipulation, databases and specification writing

Practices have shown the dominance and wide-spread use of IT industry-led software packages. For example many practices used Microsoft Suite products (Word for word processing, Excel for spreadsheets, PowerPoint for presentation, Access/Fox Pro for databases — reflecting Microsoft's 90% dominance of the software business), and Photoshop and QuarkXpress for desktop publishing. In most cases Microsoft products are often pre-loaded on to PCs when they are bought. Fewer practices still use the once popular Word Perfect word-processing software, now owned by Novell, and other word-processing software packages.

Notwithstanding the dominance of Microsoft products, such as Word or Windows 95/ NT operating systems there is a need for organizations not to become complacent but to be more vigilant, to examine options, to review strategies in the light of IT developments.

The early business applications for computer systems dealt with the payroll and accounts, characterized by the same activity each month, posting transactions to the ledger and issuing cheques. Computers could be programmed to repeat these steps over and over again. Eventually tasks that were not repetitive were automated followed by the development of expert systems that aid the thinking process.

Practices face the challenge of effectively locating, sharing, and storing the large number of different types of drawings and documents that they create. Some still use the traditional archiving and retrieving operations of scanning every document onto microfilm and manually filing it away. Others have realized the need to electronically share, view, mark-up and manage these types of documents as a necessity to achieve higher productivity by eliminating massive printing and microfilming costs and are employing systems and image viewing software packages to help improve storage and retrieval of information electronically. Typically a designer simply moves the completed CAD models to the CAD database, and at night the CAD model is automatically plotted to an HP-GL file and then, using the converter program ALLegria (gXconvert), it is automatically converted to TIFF raster files (in batches if required), and stored.

Another challenge is the never-ending search for more effective team and project management practices. Popular project management software includes PowerProject and Microsoft Project while the NBS on disk is used by the majority of practices for specification writing.

IT applications for CAD

Practices have shown the dominance and widespread use of IT industry-led software packages — AutoCAD and MicroStation. There are deficiencies in the CAD products available with much focus on drafting (86% of practices in the UK are still entrenched in the drafting mentality). These products do not reflect an integrated approach to solving problems in the design process and to addressing all the needs of the design and construction teams. Like many CAD systems, these tools were designed for the most part to improve the drawing, storage, maintainability, and clarity of architectural/engineering drawings. They were never intended to go to the head of the design process and become the primary design tools. It is madness that the process, principles and practices are driven by, and are often forced to fit, the tools, and consequently computer-aided products are seen as the cure rather than a sound design-process methodology. The important point is that practices need to understand and define their processes, principles and practices in order to choose the right tools for the job and should not be dependent on software and hardware vendors and manufacturers. It is not surprising, perhaps, that investigations into the practice of computer usage by Professor Pugh found that engineers who were working on innovatory, or design-led products, used very little software while those working on conventional, or production-led products used software extensively.[1]

The design process evolved long before computers arrived. The process is based upon understanding the specifications of a product, using mathematical principles to solve problems, developing geometries, and producing the product. Engineers, manufacturers, etc. were around before the computers or IT professionals but these have been able to leapfrog the engineering profession. Also, IT concepts come directly from engineering standards, for example, nodes, networks, reverse engineering, re-engineering, architecture, prototype, requirements and design, all come from the language of engineering. The computer tools were developed because they were good at performing simple tasks in a consistently repetitive mode — repetitive tasks with well-defined processes, practices, principles and decision points which lend themselves to automation. These tools are still primitive, based on performing two main operations, addition and subtraction with the rest of the functions involving manipulating data while sending and receiving it to storage media and devices.

From the beginning of the design process during concept development, intuitive conceptual-design computer software packages with outputs of free-form sketches and 2D illustrations are only now being used instead of sketching on tracing paper. Most designers either avoided CAD software entirely for many years — sketching ideas by hand and then passing these along to the department's support staff to render for CAD applications or to different disciplines. Drawbacks of this linear working approach were that it was time consuming and it was an iterative process that involved getting a CAD design back and then requesting changes and revisions followed by more changes which inevitably meant compromises over the time spent in getting the design right. Even those practices which applied CAD software reverted back to tried and tested methods because of its basic nature and the poor drafting and modelling tools available in the 1980s.

Traditionally practices had a lot of flexibility and the designer had control of all the tools which were readily available for the job at hand — adjusting a parallel bar or T-square; changing the angles of an adjustable triangle; use of an architectural scale for design and for layout spaces; use of an engineering scale for site layout and design. Architects/engineers have been drafting for more than 400 years to achieve this flexibility. They have automated their drafting only in the past twenty years with methods still seeking to mirror these traditional approaches, for example 2D drafting of plans, sections, elevations, and a 3D visualization, first separately and then co-ordinating them to fit together rather than representing the building

in 3D at all stages and thereby minimizing close management and constant checking. Notwithstanding all these practices, the next twenty years will see an increased use of 3D-based software to simulate buildings and expand 3D working which is currently restricted to specialized areas such as design visualization in the larger practices. The simulated buildings will create opportunities for new or enhanced architectural/engineering services such as

- computer rendering, animation and virtual reality scenes to help community groups, financial backers, prospective tenants or customers visualize the designs in 3D
- 3D facilities management, e.g. testing of building security and management staff training
- simulation and visualization of buildings' material performance
- exploration of design and maintenance alternatives
- feasibility studies of alterations
- simulation and planning of design changes required over the life of buildings
- optimization of energy use.

Examination of the case study organizations shows a variety of experiences and problems. For example, separate CAD systems have been developed for all the different members of the project team — engineers, architects, surveyors and their clients. This is an obstacle to communication and collaboration. Inter-operability, the re-working of existing CAD systems so that data can be transferred seamlessly between them offers hope in tackling this issue. Clients and project sponsors could also help by laying down briefing guidelines for the exchange and co-ordination of all information, such as specifications.

Another issue is that of outsourcing. In a number of practices there is a lack of certain computer skills, for example generating 3D models, and so this is often done externally by firms who have emerged to meet this demand. This process is similar to going to an illustrator for perspectives. Further benefits are that the computer systems are not 'clogged up by the large 3D CAD files sent for plotting' by inexperienced staff.

IT applications for telecommunications and networks

Of importance are ISDN leased communication lines which obviate the need for separate modems, intranets, the Internet and E-mail.

On-line services, such as Compuserve, the Internet, and e-mail and the adoption of laptop computers, mobile telephones, 3D graphics and virtual reality are opening lines of communication locally as well as all over the world and are changing the way architects and engineers do their jobs and run their business. Architects and engineers are able to work and interact in real-time on the same drawing across long distances at remote locations and while on site. They may also be able to take on larger projects than they could normally as the sole proprietor. High-powered laptop and note-book computers are used to make presentations to clients and to increase client involvement in the collaborative design process.

Notwithstanding this, there is a lack of understanding and mis-conception surrounding the full benefits of the Internet. For example, operating over the Internet can by-pass long-distance telephone charges and allows many user-groups access to a wealth of information, such as advice on software and hardware, e.g. facts on the latest releases and technical support, company information, background information for projects, product data, computer and high technology topics, etc. The key to filtering the most reliable, accurate, helpful items from the millions of promotional pieces and trivial facts that litter the information superhighway is to facilitate retrieval by knowing where to go by using the top search services from the 1800 plus services available on the World Wide Web, and also by knowing what kind of site contains given information. The top search services include Infoseek, Lycos, HotBot, Excite, AltaVista, Open Text.

There is a lot of hype about the Internet and how it promises improved communication, global collaboration, and tools to aid designers interact with data stored on the web, but the question was to discover how far the organizations were actually using it. A number of firms were building their own intranets (essentially a company-only Internet) and extranets (intranets that selectively allow in outsiders such as clients). An observation about some of the case study organizations examined, is an increasing interest in using the 'Project Home Page' — a web page on a private intranet that serves as the point of entry into a central site, or repository, of the project activity. The page covers information such as contacts, CAD files, site photographs, scanned photographs, reports, trans-mittals, and other types of documents. There are several organiza-tional benefits in terms of centralized communication/collabora-tion/ project management functions, for this approach.

- It enables better communication between team members with more 'contact time' and means that less critical information gets 'dropped between the cracks'.
- It enables speedier generation and processing of new information or data.
- It enables more common understanding and faster turnaround on team decision-making expediting all team members to stay up to date.
- It improves team synergy and buy-in objectives.
- It shortens product-development cycles or faster project completion and an earlier introduction to the market.
- It improves product innovation quality and value based on improved communication and effective contribution by all team members.
- It reduces development and production costs.

Table 1 is helpful in relating typical applications and some software as used at different project work stages.

IT training within the practice

The issues concern training policy, one-off training sessions on the purchase of equipment, continuing development after imple-mentation of basic skills training, for example training employees to understand the ramifications of Internet e-mail to reduce the risk of liability for the actions of employees. Training is now more important than ever as the software (e.g. CAD/CAM software and new technology concepts such as sketching, solids, assemblies, constraints, features, parametrics and bi-directional associativity) becomes more and more sophisticated and industry specific. Traditionally, vendors offer only introductory and generic training courses on the features of their system. The result of the one-off sessions is that the majority of users receive most of the actual training in an on-the-job environment, an approach which leaves many users both frustrated and unproductive.

In US CADTrain's experience, the best training occurs when

1. you have a compelling reason to learn
2. the timing of the training coincides perfectly with your needs
3. the training only addresses topics you need to know
4. the instructors go at your own pace and in a sequence that makes sense to you

Table 1. Project work stages related to examples of IT applications

Work stages: e.g. RIBA Plan of Work, Latham's Pre-project, Project and Post-project stages	Text-graphics manipulation applications: word processing, spreadsheets, DTP, databases, schedules, specifications, time management, planning
Needs identification (**Statement of Need**) Options Appraisal (Strategic Brief) Feasibility, Outline and Scheme Designs Generic/Specific	For example: • Drafting outline performance and prescriptive specification on Microsoft Word or WordPerfect or Word Pro • Floor area calculations on Microsoft Excel • Drawings/document register Fee/area calculations on Microsoft Excel • Used accounting software Sage suite of programs for accounting and job costing
Design development (**Project and Detailed Briefs**) Final Design Client and Statutory Approvals: submissions for planning, building regulations Board approvals, corporate design policy	• Specification prepared as table with cost plan on Microsoft Word and Excel • Building Regulations and Planning Permission reports on Microsoft Word • Project architects/engineers type minutes on Microsoft Word • Develop material for Client presentation using Quark Xpress and Microsoft Powerpoint • Specification fixed using NBS and worked up using Specification Manager
Management of project resources (teams, money, etc.) Manage Time (program) Manage Risk Manage Process (project planning and administration)	• Preliminary programs on Microsoft Excel and Microsoft Planner • Register set up for subcontractors' drawings on Microsoft Excel • Design standard forms on Microsoft Word and Excel • Used Document Management System TeamMate as electronic drawing storage and retrieval system after consideration of Microsoft Access, Claris's FileMaker Pro, Corel's Quattro Pro • Use Asta Power Project for project management software after looking at Microsoft Project and CA-Super Project
Project Completion	• Snagging sheets prepared as table on Microsoft Excel • Specification developed as a maintenance user manual on Microsoft Word and Excel
Commissioning and handover (Commissioning Brief)	• Display Notices and Instructions developed on Microsoft Publisher
Occupation/Use (Facility Management Brief) Post Completion and Evaluation Reports (**Feedback Review of 3Ps** — Process, Performance and Product)	• Service contract documentation and Health and Safety Instruction records on Microsoft Word • Prepare weekly and monthly Management reports on Microsoft Word

Scientific/mathematical/ financial/other specialist applications: structural, energy, cost, etc. analyses (using Hevacomp etc.)	CAD/CAM applications: 2D and 3D drafting, visualization (AutoCAD, MicroStation, SpiritCAD, MiniCAD, ArchiCAD Form Z, Powerdraft, etc.)	Telecommunications and Networks: ISDN links, Intranets, Internet (meetings, conferences, negotiations, etc.)
For example: • Used accounting software Sage suite of programs for accounting and job costing • Consulted RIBA IT Construction Information Service (CD-ROM of Technical Indexes and RIBA Services)	**For example:** • A3 design studies to demonstrate most efficient way of fitting complex job on the irregular site • Floor plan developed and inserted on AutoCAD CAD survey • Preparation of optional elevations on AutoCAD	**For example:** • Client/design meetings and communications discussing business and constructions options • Negotiations with planners • Meetings with interest groups, press and media
• Building Regulations calculations for ventilation and heat loss using proprietary software • Structural calculations using in-house spreadsheets • Used SPSS for Windows for (cluster) statistical analysis of space areas • Used Radiance program to trace light from natural and artificial sources to generate a full 3D image of the illuminated main space	• Prepared first 3D model using Accurender 3D Studio a *de facto* std rendering animation package • Elevations developed with preliminary 3D model. CAD facilitates quick revisions, accurate drawings and 3D design studies • Drawings developed and annotated for submissions. The Building Regulations set formed the basis for the working drawings (building regulations information on frozen layer)	• Perspectives presented to client which in turn releases more money for 'enhancement'. Photoshop and Corel Draw to tweak CAD models • ISDN link established between Client and Design Team members to facilitate effective communication exchange of large files, such as drawings. Considered using www.ukindex.co.uk for a mailing list
• Used Timesheet Expert for Windows to electronically record time and expenses • Further CFD analysis • Used Hevacomp (mechanical and electrical software) by Smithwood House for duct sizing in order to validate information from schedules and schematics, i.e. line drawings showing electrical distribution cables, pipes and ventilation ducting • Simulation studies and verification of results using Wind Tunnel	• Minimum number of files produced. Masterfile allows the same drawings to be used for scales. Engineers (Structural, Building Services) issued with base files. • Full 3D model prepared and used for full render and for hand drawn views. Client had problems with materials so colour 3D images prepared • Drawings modified as necessary to save costs and to clarify details. Rendered view of core areas prepared for client agreement. Later exhibited at RIBA • Site Architects/Engineers equipped with laptops and AutoCAD LT	• Used Lotus Agenda for paper-based appointment books — storing names and telephone numbers, taking notes, etc. • Used E-mail for quick communication with members of the project team • Selected project information made available on the practice Novelle Netware intranet to allow easy sharing of files on the office high-powered server computer by project team members
• Transferred Excel data to StartView (statistical software distributed by Cherwell Scientific) for further analysis	• 3D computer models with photograph images of physical building	• ISDN links used to communicate with Client and Design Team members about completion issues
• Checking Building Management System (BMS) for computer controlled environments to optimize HVAC efficiency and equipment safety	• Preparation of maintenance drawings showing location of emergency call points and lighting plus service inlets and hydrants • Photos of completed project in practice brochures	• ISDN links heavily used for arranging handover and to obtain client approval of brochure for the handover party
• Logging and tracking of performance from contractors	• As built or turn out drawings saved as A3 rather than A1 as clarity is sufficient for 3D building data. CAFM (Computer Aided Facility Management) systems used to track inventory of equipment furniture and appliances repair schedules, rebuilding model using Archibus FM. This is in order to keep a tight grip on cost	• Consulted *Construction Law Online* for the latest contract law decisions on potential contractual problem

5. the training follows a structured learning methodology

6. you work with your products.

    Also four levels of learning are identified.

-   Level 1: awareness, familiarization, recognition.
-   Level 2: understanding, discrimination among alternatives.
-   Level 3: experience, skill in application.
-   Level 4: integration of knowledge, full competency.

Practices in this publication recognize the need to train their staff (designers, engineers, analysts, technicians, draftsmen, etc.) but face the dilemma as to whether this is best done in-house or externally at training centres or at the dealer's premises. There may also be difficulties with a training budget, for example a poor course may be run externally that is not value for money or there is no time to do the training if an office is busy and has to meet tight deadlines for projects. Sometimes, and especially with older staff, there is a reluctance to learn computer skills. Once trained, staff need to spend several hours a week working on the product to keep up-to-date. Universities could help by training students not only with the basics but also teaching them that practices have different CAD packages and that these are up-dated frequently in some practices. The importance of CAD skills is increasing, with demand rising and local colleges are now offering CAD courses ranging from City & Guilds (AutoCad and Microstation) spread over a year to intensive one-week courses.

Issues of concern to the practices are the same as those for the IT industry as a whole — e.g. use of illegal software and the legal aspects of replacing printed communication with electronic mediums such as e-mail.

Spending on IT has become the second biggest item on the budget for most organizations, but it's almost impossible to measure returns. Come to that, it's almost impossible to measure the investment.[2]

The problems are the difficulties of measuring the intangible costs of IT infrastructure, the intangibles (downtime, network failures — 91% of companies suffer network failures every year). Additional charges on telephone costs, printer toner, etc. in a situation in which IT assets and benefits, including both software and hardware, are simply unknown or not easily monitored and tracked.

In desktop hardware operating systems — networks and servers — increase in base computing (aided by parallel processing) and graphics performance are likely. Video and multi-media applications will benefit from the likely developments in Data Compression hardware with increases in the capacity of DRAM (Dynamic Random Access Memory) and improvements of bus structures. The increasing convergence of several related technologies including computing, communications, consumer electronics, entertainment, telecommunications and broadcasting, all of which are based on the core technology of microelectronics, signal a turbulent mixing of the underlying technology giving rise to several discontinuous changes in these areas.

The area of networking and telecommunications encompasses all the matter in the various chapters of this book. The growth of client/server architecture, high performance databases, parallel processing hardware, text and multi-media retrieval tools all serve to enhance on-line information providing services. The future may be based around surfing the Internet, multi-media CD-ROM, VOD (video on demand) and VR.[3] There are three main advantages for web-based technology.

1. A web-based system will be cheaper to maintain, largely because software upgrades or tweaks are performed only on the central server rather than on every PC running the package. This cuts down the operating costs and saves time.

2. A web-based system is designed for remote access. Employees anywhere in the world can access simply and efficiently up-to-date company information or databases via a laptop or even using a hand-held organizer. Access to important information no longer depends on a cumbersome PC directly linked to a firm's server. This is particularly useful for industries that demand a lot of off-site work, such as construction. For example, by making all relevant project information available on-line, web-technology reduces the amount of documentation that needs distributing throughout a project. CAD stations can view files very quickly without the overhead of a full workstand. Possible scenarios cover all members of a construction project, from designers through to the finishing trades communicating electronically at very high levels of detail using the latest multi-media software. Also, these scenarios may involve designers wanting to know immediately

what designs a window manufacturer produces, or designers urgently requiring a project-approved detail for drawings or even a site engineer wanting to have a close look at the latest foundation detailing on a drawing or sending and receiving drawings via E-mail. 'A problem in one place may already have been solved elsewhere in the world. Electronic communications connect experts worldwide to capture their knowledge and the details of who-knows-what in databases.' (Skyme 1998)[4]

3. Web-based software will revolutionize EDI links to suppliers. When a firm's head has one version of EDI and their supplier has another, vital orders lie unprocessed because the systems cannot communicate. The web uses well understood and genuinely open 'protocols' — which means that every web-enabled system can communicate easily with every other one, regardless of hardware platform or (usually) operating system.

Advances in human-computer interface (HCI), HCI-related developments and speech technology should have an impact. Improvements are also expected in terms of visualization to virtual reality.

Nigel Gilbert makes a predictive analysis on the effect of VR on architectural practice in the 21st century in which cyberspace of computer graphics systems like Quantel Paintwork anticipates architecture's passage from the formal 'natural relations' of Norberg Schultz to the endless thin entries of Jencks 'condition zero'.[5] VR will not only be empowered with voice response and synthesis but also with display capabilities now only present in the computer dreams — the world of full-time simulation.

The 'cyber' space of electronic communications is the simultaneous space in which our lives are real and exist dimensionless and locationless in parallel with 3D visual space.

Alan Davidson of Hayes Davidson talks of 'virtual architecture' as one which is not necessarily analogous to real structures and one which could perhaps spawn real architecture.[6]

Yoav Etiel of Bentley Systems (software developers of Microstation) speaks of another facet of the model for the future.[7]

In order to realize the next big leap in productivity, we must shift the architect's and engineer's focus from an electronic drawing to the digital, single building model . . . Since Java provides for a consistent GUI in both thick and thin clients, it saves training costs, along with cost of documentation. It reduces cost of software maintenance and installation thanks to auto-install and auto-update features, and provides application-level inter-operability with enterprise IT. In short, Java is the right option that arrived at the right time — just when organizations have shifted their focus to enterprise-wide collaboration and project life-cycles . . . Infrastructure is red hot, and infrastructure projects drive much of our users' businesses. With so much business to win, technology must help, win, deliver and support such projects. There is also a lot to lose: with economic borders literally wide open virtually all over the world, architects and engineers are going to either compete or team with global organizations on these projects, while at the same time seeking projects in remote locations in order to fuel their own growth.

References

1. Pugh S. Creating innovative products using total design. D. and R. Andrade (eds). Addison Wesley, 1996.

2. Young R. IT strategy: what's this lot worth, then? *Financial Director.* June, 1998, 38.

3. Negroponte N. *Bring digital.* Hodder and Stoughton. 1995

4. British Research Establishment. *Information Technology trends and the construction industry.* BR (British Report) 269. BRE, Watford, 1994

5. Gilbert N. Condition Zero. *World Architecture,* No. 26, 1993, 78–81.

6. Davidson A. My computers: uses computer graphics to represent architectural work. *Architects Journal,* 13th Oct., 1994, 7.

7. Sprohrer R. Embracing progress. *World Architecture.* No. 71, Nov. 1998.

Glossary

A

Accumulator refers to the central register in a processor's arithmetic and logic unit, the place where in effect all the processing takes place; and thus the major bottleneck in a system that does not use parallel processing.

Active X Plug-ins used with the browser Microsoft Internet Explorer.

Add-in and **Add-on** elements to be added in such as an auxiliary card or drive internal to a PC, e.g. internal modem, fax card, VGA display card, disk drive, etc. and elements to be added on such as ancillary equipment external to the PC, e.g. printer, modem, scanner, etc.

Address refers to

- in communications: a coded representation added to a block of data in transfer (e.g. a packet in a packet switching system or E-mail message) to show where it should go (destination)

- in computing: the unique number that allows access (location) to any given storage or screen (picture) cell or disk sector.

Algorithm refers to a set of well-defined rules or procedures (sequence of steps), typically composed of mathematical functions or geometrical functions, for performing a task or for solving a problem in a finite number of steps such as removing hidden lines from the display of a 3D solid object. It is also a proof that the problem has no solution. Ways of expressing algorithms include

- (*a*) clear English sentences

- (*b*) PSEUDO-CODE, somewhat like program code (indeed, a program is an algorithm itself)

- (*c*) a STRUCTURE CHART

- (*d*) a FLOW CHART.

The main advantage of this kind of problem-solving system for computing purposes, is that it reduces the problem to a series of yes/no options which are easily translated into binary form. Compare with heuristic (pertaining to exploratory methods of problem solving in which solutions are discovered by evaluation of the progress made toward the final result 'self teaching' or 'trial and error').

Analog(ue) The use of physical variables such as distance or rotation to represent and correspond with numerical variables occurring in a composition. In numerical control, the term applies to a system that uses electrical voltage magnitudes or ratios to represent physical axis positions. Analog(ue) representation of data is mainly used in an analog(ue) computer. Compare with digital representation of information.

Anthropomorphic an adjective with the literal meaning 'human shape'. An anthropomorphic robot is one that looks more or less like a human being.

Application Application program-software that provides services such as electronic mail or file transfers. Applications software refers to programs used for specific tasks (e.g. payroll, writing letters, keeping records, producing drawings and charts, information retrieval, data analysis) as opposed to general-purpose systems software that manages and controls the working of the computer and carries out system tasks such as formatting disks, reading files on the disk, controlling the screen display and so on. Systems software is made up of the operating system, the operating environment and utilities. The operating system controls the disk drives and other hardware devices, including the network operations in some cases allowing the computer to do housekeeping tasks such as loading, copying and deleting files. The operating environment provides an easy-to-use way of carrying out housekeeping tasks as well as providing a friendly environment within which computer applications can run. Examples are the Apple Macintosh operating environment and the almost identical GEM (Graphics Environment Manager) from Digital Research and WINDOWS from Microsoft or Presentation Manager from IBM all of which are based on the earlier operating environment developed by Xerox at the end of the 1970s and popularized by Apple Macintosh. Utilities, often referred to as front ends (to the operating system because they do not allow running of applications within this type of software), provide additional facilities and routines beyond the capabilities of the operating system or carry out operating system tasks in a more efficient and easier manner. Examples are PowerMenu, PC Tools, Qdos, and Xtree to DOS.

Architecture The physical and logical structure and therefore the overall design of a computer or communications network and the way in which the components relate to one another.

Arithmetic Logic Unit (ALU) The part of the computer processing section that does the adding, subtracting, multiplying, dividing and/or logical tasks (comparing). The accumulator is a register, close to a computer's arithmetic logic unit, in which the results of arithmetic and logical operations are calculated. The results are transferred from the accumulator to memory pending further operations or for storage.

Artificial Intelligence (AI) is the ability of a computer to reason and learn as a human being does. It includes expert systems and usually a set of three programs — 'knowledge base', an input program and 'inference engine' — written to solve problems or make decisions in a specified field) and 'knowledge engineering'. The term was first coined by MIT's John McCarthy (author of the programming language LISP) in the 1950s. The two main areas of research are

- (*a*) an investigation into the nature of knowledge and the use of it in decision making

- (*b*) using the results of the first to develop computer systems capable of demonstrating intelligence so that they may not only make decisions based on their current state of knowledge, but also add to their knowledge bases as a result of their experiences.

ASCII (American Standard Code for Information Interchange) pronounced 'asskey' — 128 characters used throughout. An ASCII text file can be printed, but no extra (formatting) characters are included. Also the first characters of the 255 IBM character set (known as the extended ASCII character set) — 'The IBM character set'.

Automatic Dimensioning is a CAD capability that computes the dimensions in a displayed design, or in a designated section, and automatically places dimensions, dimensional lines and arrowheads where required. In the case of mapping, this capability labels the linear feature with length and azimuth. Azimuth is the direction of a straight line to a point in a horizontal plane, expressed as the angular distance from a reference line, such as the observer's line of view.

Automation is the use of technology to accomplish routine, repetitive, non-creative and one-of-a-kind activities.

B

Background processing refers to

    (*a*) in computing, the execution of lower priority computer programs when higher priority programs do not require any system resource

    (*b*) in word processing, the execution of a user's request such as printing a document whilst the user is performing other tasks.

Back-up The process of making up a copy of a disk file or files just in case the original is corrupted.

Bandwidth in telecommunications, refers to the extent of the frequency spectrum within which signals such as those carrying information can pass through a system without significant attenuation. Bandwidth will have to increase dramatically before we are able to access instant libraries of books, audio recordings or movies on the Internet.

Batch in computing terms is

    (*a*) an accumulation of data to be processed

    (*b*) a group of records or data processing jobs brought together for processing or transmission.

Batch processing The processing of data where a number of similar input items are grouped or accumulated as a single unit (batch) for processing during the same machine run. (The processing of transactions a group at a time instead of as they arise, for example banks update customer account balances regularly at the end of each day in one batch rather than contin-uously and interactively throughout each working day.) Batch processing is the main mode of operation of main frame systems; the operating software includes a JOB CONTROL LANGUAGE (JCL) for this purpose. A JCL program (command file) will tell the system such things as

- the code for the job's owner (i.e. whom to bill)
- the job name
- the program language used
- the drivers that contain the program and data files
- the printer to use
- the job's priority. It is useful for control purposes but involves a delay between an event (e.g. removing an item from store) and its appearance in the computer records. Compare with 'Background processing'. Batch processing also refers to the technique of executing a set of computer programs such that each is completed before the next program is started.

Baud or baud rate The rate of data transmission over a communications link, i.e. a measure of signalling speed in a digital communication circuit. The speed in bauds is equal to the number of discrete conditions or signal

events per second. The name comes from that of Emil Baudot, who devised the first standard telegraph code in 1887. Twelve hundred baud is approximately 100 characters per second. (Compare with RS-232.)

In the computer world the main communications standard is the RS-232 defining the connections to be used in the cabling that links computers. It specifies a channel for transmitting data, a channel for receiving data, and channels for control signals. It also specifies that data is transmitted serially, i.e. one bit at a time, and so the RS-232 socket on the computer is often known as the serial point. The RS-232 standard does not specify a single baud rate, but it allows a number of rates: 75, 150, 300, 600 and up to 9600 and above. Commonly used rates are 300 and 1200 and many on-line systems (such as Bulletin Boards) send and receive data at either 300 or 1200 baud. One exception, though, is Prestel — when connected to this, you send data from your computer at 75 baud, and receive it from the Prestel computer at 1200 baud. The communications software normally sets the baud rate before the computer can be used.

Binary Number System uses the base 2 as opposed to the decimal number system which uses the base 10. The binary system is comparable to the decimal system in using the concepts of absolute value and positional value. The difference is that the binary numbering system employs only two absolute values, 0 and 1. Because there are only two digits; the positional significance of a binary number is based on the progression of powers of 2. The numbers are expressed in binary notation as series of 0s and 1s commonly referred to as bits. The 0 is described as no bit and represents an 'off' position. The 1 is described as a bit and represents an 'on' condition. The lowest-order position in the binary system is called 1-bit. The next position is called 2-bit; the next 4-bit; the next, 8-bit and so on.

BIOS A collection of software codes built into a PC that handle some of the fundamental tasks of sending data from one part of the computer to another.

Bit A contraction of '**B**inary dig**IT**' — the smallest unit of information in a binary number system. Eight digits are needed to create one byte or character. A bit is the smallest entity of a memory word in which a value can be stored. In the binary system, a bit can represent either 0 or 1; to a computer, a bit will indicate an off or on signal. Bits are the units of information that, when combined in certain configurations, will signal to the computer what it is to do. A bit may assume only one of two values 0 or 1 (i.e. On/Off, or Yes/No). Bits are organized into larger units called words for access by computer instructions. Computers are often categorized by word size in bits, i.e. the maximum word size that can be processed as a unit during an instruction cycle (such as 16-bit computers or 32-bit computers). The number of bits in a word is an indication of the processing power of the system, especially for calculations or for high-precision data.

Black box A device that performs a specific function, but whose detailed operation is not known, or not specified, in the context of the discussion.

Block refers to an amalgam of computer records, handled for efficiency as one unit for purposes of storage, input and output.

Boolean Algebra A process of reasoning or a deductive system of theorems using a symbolic logic and dealing with classes, propositions, or on-off circuit elements such as AND, OR, NOT, EXCEPT, IF, THEN, to permit math-ematical calculations. Algebraic or symbolic logic formulas (adapted from the English mathematician George Boole's work 1815–1864) are used in CAD to expand design-rules, check programs and expedite the construction of geometric figures. The logical expression and analysis is usually illustrated by means of a truth table such as

X	Y	Z
0	0	0
1	0	1
0	1	1
1	1	0

'If X may be true or false, and Y may similarly be true or false, which combinations of truth and falsehood of X and Y lead to a true or false Z?' In the above example 0 corresponds to 'false' and 1 corresponds to 'true'. It can be seen that the Z postulated in this relationship is true if one of X or Y is true, but not if both X and Y are true or false.

In a computer-aided-mapping environment, Boolean operations are used either singly or in combination to identify features by their nongraphic properties (i.e. by highlighting all parcels with an area greater than 15 km² etc.).

Browser Software that allows viewers from any platform to view pages in HTML format. Popular examples are Netscape Navigator and Microsoft Internet Explorer. Netscape is the most popular browser (with, in 1997, about 85% of the market share) representing also the world's most widely installed application.

Bug refers to an error in the program, a software error. Also used generally to express any computer malfunction. Debugging is finding and eliminating a bug from a computer program or system, hence taking corrective action.

Bureau A firm that provides IT services (such as running of the monthly payroll programs) for others, e.g. FAX BUREAU, COMPUTER BUREAU.

Business process re-engineering refers to the dramatic organizational changes made to exploit the competitive advantages of IT.

Business Systems A general term used to describe the various types of equipment and operating routines which, brought together into some form of network, provide a means of gathering, storing and redistributing business information. Associated terms include the electronic office, automated office, business automation and telematics. Local area networks (LANs) are computer-based business systems.

Byte A cluster of 8 bits used to present one character in most computer systems. Hence usually used as a synonym for character when defining capacity of storage. A measure of the memory capacity of a system, or of an individual storage unit (as 300-million byte disk). There are 256 possible words in a byte, e.g.

One byte

Binary (8 bits)	Decimal	Hexadecimal
00000000	0	0
00000010	2	2
00001010	10	10
00001011	11	a
00010000	16	f
00010001	17	f1
11111111	255	ff

In the base 16 number system the (16) hex digits are 0 1 2 3 4 5 6 7 8 9 A B C D E F; the F has the same value as denary ('decimal' 15; 16 is 10 in hex). A hex number is a set of hex digits, the normal rules of place value making these worth units (16 to the power 0), sixteens (16 1), 256s

(16 2) . . . moving left from the hex point, and 1/16 (16−1), 1/256 (1−2), 1/4096 (16−3) . . . moving right from it. Thus the hex number 4F.A3 has the same value as denary 79.6367. . .

C

CAD (Computer-Aided Design) The use of a computer program to aid design which involves computer graphics, modelling, analysis, simulation and optimization of designs for production. The computer program is used to generate designs normally in the form of dimensional drawings. CAD uses the Cartesian co-ordinate system (a set of numbers defining the location of a point within a rectilinear co-ordinate system consisting of three perpendicular axes X,Y,Z) defined by René Descartes, the 17th century French mathematician and philosopher, to allow representation of objects in the 2D and the 3D world. Therefore the system allows one to draw lines, arcs and circles as geometric entities on the computer screen which can then be dimensioned to output the appropriate drawing on a pen plotter or similar device. The CAD system builds up a database of various entities created and can then store them on computer disk for future use or modification. There are many benefits of using CAD systems for example one can make design changes easily and because the design is stored electronically there are none of the problems associated with physical storage. Bentley Systems and Autodesk are clear market leaders. Others are CADlogic, Leonardo Computer Systems, CODEC, and so on.

CAM (Computer-Aided Manufacture) Although CAM is often used to define computer-aided manufacture the term can be all-encompassing to include production, scheduling, and the like. More often than not it actually refers to computer-aided machining.

Cache A high speed store situated logically, if not physically between a processor and its main store. Disk cache (memory-cache — processor cache) is a PC performance enhancement technique (usually implemented by a configuration file or program, but can also be hardware) which uses extra memory to provide fast access to predictable data needed by a program or processor, e.g. SMARTDRV.SYS

Cathode ray tube (CRT) A device in electronics that converts electrical signals to a visual display. It comprises an evacuated glass envelope shaped as a television tube, an electron gun, focusing deflection systems and a screen coated with phosphors.

CD-ROM (Compact Disk Read Only Memory) is a compact disk for looking at only rather than re-storage. It is like an audio compact disk except that it stores vast amounts of information (a mixture of text, data, pictures, film voice and music) rather than sound. It is used for databases of information such as encyclopaedias. CD-ROMs offer the potential of so much capacity for their physical size and are convenient and cut down on cost of distribution especially for heavy weight applications. However, the slow transfer speed of data from the CD-ROM is one drawback, another is that to write data on CD-ROMs requires an expensive CD-ROM maker. Even then it is necessary to use a special type of CD disk known as WORM (WriteOnce, Read only Many times) on which you only write once — and therefore it is not re-recordable.

Cell

- a collection of related geometric and/or alphanumeric data on one or more layers of a design on the system

- the physical location or design of a single data bit in memory

- a collection of lines, curves, surfaces, lettering, dimension lines and possibly other cells, all of which form a standard symbol or a sub-assembly in a design or drawing.

Central Processing Unit (CPU) refers to the silicon chip that manipulates data within the computer, defined according to speed in ascending order 286 80286...386 80386...486 80486... and the Pentium from Intel effectively the '80586', the fastest microprocessor. It has been succeeded by the next generation Intel's processor technology coded named 'P6' (a semi-conductor component or group of components containing extensive circuits and which implement the central processor of the computer). The P6 is reputed to be especially well-suited for upcoming desktop applications like speech recognition and multi-media authoring, as well as for more demanding server and workstation applications. Also this is the 'heart' of the machine. Cut down versions of the microprocessors are SX chips and those more powerful are DX chips such that 80486SX is slower than the 80486DX. Early IBM PCs named XTs were based upon an 8088 microprocessor with later machines using the faster 8086 processor chips. With this in mind the main areas of hardware specification are CPU type (microprocessor, e.g. 386, 486 or Pentium), CPU speed (measured in millions of cycles per second — MHz or megahertz, e.g. 20 or 25 MHz for the 386 chip, clock speed of 30 or 35 MHz for the 486), RAM capacity and graphics format (such as MDA, CGA, EGA, VGA, MCGA, PGA or XGA display formats).

Centronics connector port (interface between PC and other device such as a printer or modem consisting of plug, socket and communications specification) for the parallel data communications standard, a 36-way plug and socket (30 wires used) at the printer end of the cable.

Character An alphabetical, numerical or special graphic symbol used as a part of the organization, control or representation of CAD/CAM data.

Classification and Coding systems Conventions which provide a logical and meaningful basis upon which to code information or artefacts. The main aim is to identify items in a way that facilitates easy identification and selective access and retrieval (e.g. the Brisch, Dewey Decimal, NATO, Optiz, PERA and Pittler systems, etc.). The Dewey Decimal library classification, a system devised by Melvile Dewey to classify areas of knowledge, is still used in modified form. In this system there are ten main numbered classes (e.g. Philosophy = 100 and each area is sub-divided progressively into ten sub-classes and so on). In faceted classification, elements such as a book are identified so that they relate to the requirements of the person seeking the information. (See Layer for CI/sfB classification system.)

Client–server is a form of networked computing where programs on server computers can be activated by requests from client workstations. The 'client' is a PC and the 'server' is a processor where communal databases and programs are stored, all connected to a LAN. The main advantage of client–server over mainframe systems is its flexibility enabling companies to respond quickly to changes within the organization and business environment. Client–server describes an application design methodology and mechanism for data distribution. Client–server systems can support multiple 'hosts' for different applications or data-tables.

Compiler The software which translates a high-level language program into machine code to allow the system to run it. The compiler checks for, and reports, any syntax errors in the source program which, if free of errors, then produces a complete object code program.

Computer graphics The use of computers to generate and display pictorial images. A user can generate these images using either a keyboard, or some special graphic input device.

Computer memory An area of storage within the computer where programs and data are held ready for processing by the CPU (central processing unit) and therefore can be readily accessed at extremely high speed unlike disk storage. This memory is of two types: RAM (temporary store for holding programs and data loaded from disk, or typed at the keyboard or input from some device) and ROM (permanent store for holding programs) (see definitions).

Contiguous file A disk file (program or data file) which occupies one area on a diskette or hard-disk, as opposed to a non-contiguous (fragmented) file which occupies more than one area on a disk.

Cursor surrogates The conventional cursor (line or blob) that is used on screens to define a position is too slow for some purposes, and several alternatives have been produced, one of these is the light pen. This is a wand that can be placed on a screen to pinpoint a position that one wishes to indicate to the computer. Similar is the touch screen. This is a screen that is sensitive to being touched by one's fingers. This is widely used where frequent telephone-based communications are made (such as in a bank dealing room). The screen is filled with the names or codes of contacts; by touching the contact name, the computer will automatically dial the correct number.

D

Data

- Information (numeric or alphanumeric, etc.) which is to be input, processed in some way and output by the computer

- a representation of facts, concepts, or instructions in a formalized manner in order that it may be communicated, interpreted, or processed by human or automatic means.

Data provide the building block for information. Data communication is the transmission of information in digital form using communications networks and equipment especially designed for this purpose.

Database

- A general term for any large body of data, usually relating to a specific area of interest

- a file of data held in a computer and structured in such a way that applications can access the data and update it without constraining its design.

Database management refers to a system of file organization enabling data to be retrieved in the manner, style and format desired (distinguish between 'database', 'relational database', 'relational database management', etc.). Databank is usually an alternative term for database and is sometimes used to refer exclusively to a collection of factual, or numerical, data, as distinct from bibliographic database, which gives references to documents.

Data Processing (DP) includes all clerical, arithmetical and logical operations on data. Data processing in the context of information technology always implies the use of a computer for these operations.

Data Warehousing Data warehouse architecture is a mechanism for bringing different data sources together through a set of client and server side tools as well as enabling services.

Default is the predetermined value of a parameter required in a task or operation. It is automatically supplied by the systems whenever that value (i.e. text, weight, or grid size) or device to receive the data is not specified and therefore requires no action from the user.

Digital refers to displaying information as numbers and being able to take only a restricted number of values between the bottom and top of the permitted range. (Compare with Analog(ue) measure which can have any value in the range or continuously changing function.)

Digitizers The term refers to devices used for reading drawings (such as maps), converting the information from analog input into digital form and then storing the result. Three of the most popular digitizing technologies are

(a) electrostatic (capacitive) which uses an electric field that radiates from the tablet surface and is picked up by a pen or cursor

(b) electromagnetic which uses an electromagnetic coupling between the cursor and the tablet to determine cursor location

(c) magnetostrictive which incorporates a grid of regularly spaced wires beneath the digitizing surface.

DOS (Disk Operating System) The set of programs and system configuration files which control the basic functions of an IBM (PC-DOS). The AUTOEXE.BAT file is a DOS batch file which if it exists must be the root directory of the DOS start disk. It executes (load, run or start) DOS commands automatically, after a PC is switched on. The CONFIG.SYS file is a DOS system configuration file which must be in the root directory of the DOS start disk. Commands, established by the user and/or by programs, which set up the PC, its options and its programs to operate properly.

Downsizing A term which refers to the increasing trend of employing small but equally powerful computers and local area networks instead of larger and more expensive mini or mainframe computers connected to dumb terminals which are more expensive to support and maintain.

Duplex A method of communication between two terminals which allows both to transmit simultaneously and independently.

E

EDI (Electronic data interchange) ranges from ordering and billing over secure data lines to full exchange of CAD models, cost estimates and project control data.

Electronic mail (e-mail) A system of point-to-point or person-to-person messages via terminals. Messages are entered (usually at a keyboard) by a sender, transmitted electronically and read when convenient by the recipient on a visual display unit. Sending e-mail involves the following.

- E-mail *client software* (Internal mail, Pegasus, etc.) contacts the Internet service provider's computer *server* (software called SMTP) over a modem or network connection and informs it that a message is to be sent to a certain address. The server acknowledges with a message, either 'Send it now' or 'Too busy: send later'.

- The SMTP server asks another piece of software, a *domain name server*, how to route the message through the Internet. The domain name server searches for the domain name — the part of the address after the '@' character — to locate the recipient's e-mail/server.

- The e-mail message travels through Internet routers (which decide which electronic pathway to send the e-mail), passes through gateways (which translate data from one type of computer system to another, e.g. from Windows, UNIX, Macintosh, etc.) and on to the next pass-through computer system until the e-mail arrives at the recipient's SMTP server (which transfers the message to the POP (Post Office Protocol) server. The POP server holds the message until it is retrieved.

Ethernet refers to a common type of local network developed by Rank Xerox, using coaxial cabling as the connection link between computers, word processors, complete workstations, etc.). Coaxial cabling is a form of high-quality network wiring that can carry data over relatively long distances across LANs.

F

Facsimile

- a system of still picture transmission and reception, using synchronized scanning at the transmitter and receiver, and telegraphy types of modulation such as frequency shift keying. The reconstructed image at the receiving station is either duplicated on paper or film

- a precise reproduction of an original document

- a hard copy reproduction.

Facsimile telegraphy (fax) or facsimile transfer is a method of sending and receiving graphics and text over a telecommunications network, by converting visual images into electrical signals and then converting them back again into a copy of the original. Fax systems (cards and machines) come in groups: groups 1 (analog) and 2 (digital) are no longer made, while group 3 (also digital, and able to communicate with group 2 systems) is the norm.

Fibre optics refers to the science and technology of carrying information or data on light waves, often in a highly transparent glass cable. A single light wave travels through the fibre by a series of internal reflections.

Flowcharting A technique for representing a succession of events by means of lines (indicating interconnections) linking symbols (indicating events, or processes). There are two main types

(a) systems flowcharts, which aim to show the relationship between events in data processing systems

(b) program flowcharts, which aim to break down a problem into logical components which are analysable by programming commands.

Font or fount refers to the complete set of characters (including spaces) of a given typeface (style, design) in a given size.

Format The predetermined arrangement of characters, fields, lines, page numbers, punctuation marks, and so on. Refers to input, output and files.

ftp (File Transfer Protocol) lets users download or upload files over the Internet (unadorned by the graphics and pizazz of a web page).

G

Giga (symbol G) is the unit prefix for 10 to the power 9, a thousand million, e.g. gigabytes (GB) for a video disk.

Gopher is an interactive file server system that offers menu-based capabilities for browsing Internet resources.

Graphics and display formats Graphic facilities are provided by the addition of a graphics adapter card which can be plugged into one of the expansion slots inside a computer's system unit casing. MDA (Monochrome Display Adapter) for monochrome text only, a resolution of 720 dots horizontally and 350 dots vertically; CGA (Colour Graphics Adapter) for games and educational use are not recommended for business use, displays up to 16 colours with the highest graphics resolution of 320 x 200 pixels; Hercules or Hercules-Compatible Adapter and Monochrome — high resolution, high quality monochrome display for both text and graphics, graphics resolution 720 x 348 pixels; EGA (Enhanced Graphics Adapter) displays both text and graphics in up to 16 colours, highest graphics resolution 640 x 400 pixels; VGA (Video Graphics Array) technically superior to EGA displays high resolution text and graphics display with up to 256 colours on screen simultaneously, highest graphics resolution 640 x 480 pixels; SVGA (Super Video Graphics Array) displays 256 colours from a palette of 16.7 million, highest graphics resolution 800 x 680 pixels.

Graphic tablets are precision input devices for manually creating images on screen normally using a stylus on a tablet of a grid of wires. They provide the electronic analogue of the drawing boards which used to be the distinguishing feature of drawing offices. Three advantages which graphic tablets (usually A4, A3 or A2 in size drawn with special stylus) have over a mouse are

(*a*) a choice of pointing devices — a puck is essential for accurate tracing or digitizing, while a stylus is better for producing rough sketches

(*b*) absolute precision and repositioning of the pointer

(*c*) improved speed and productivity.

Groupware is a term for software products that enable co-ordination between team members.

H

Hard copy Output in a permanent physical form (such as printing on paper).

Holography is the creation of 3-Dimensional images of objects using light produced by lasers.

Host A computer that controls communications between other computers or terminals on a local area network.

Hot desking is a method of organizing workspace so that each desk has all the facilities required by any employee, but no desk is owned by a single individual. A free desk can be used by any employee. Hence hot desking refers to free-address desking or notion of 'location independence' where no office worker has territorial rights over a particular desk or single function workstation. It frees-up expensive office floor space stimulating flexibility and improving interaction. Hence, it implies all round upgrading — larger desks, more meeting spaces and better designed workstations to compensate for the absence of individualized desks.

HTML (Hypertext Markup Language) is the standard coding to create content for the web.

Human–Computer Interaction (HCI) is a term which refers to a practical technique for improving the human computer interface. HCI is concerned with the design implementation and evaluation of usable computer systems. Usability covers fitness for purpose and impact on people as well as ease of use.

Hypermedia Any computer-controlled system which allows the user to interact with a number of media, e.g. text and graphics on film, on CD-ROM, and on magnetic disk. Interaction should be very flexible, so users pass through any part of the material by any paths they wish.

Hypertext A linking of related information so that one can leap from topic to topic and find related material in a database (unlike, say, word processors which use information in a linear manner). Hypertext is a simple type of hypermedia giving access to text (and perhaps very simple graphics). It creates a network enabling cross-referencing and other related material to be accessed automatically. Access can be to any digitized material: text, graphics, video, audio. The concept was originally initiated in 1965 by Ted Nelson. Hypertext software allows users to browse documents at various layers of detail, for example construction contracts according to section headings, paragraph headings, sub-paragraph headings and main text.[1]

Hz Hertz, a unit of frequency equal to one cycle per second.

I

Icon A graphic symbol of an action or object.

Information Technology (IT) refers to the acquisition, processing, storage and dissemination of vocal, pictorial, textual and numerical information by means of computers and telecommunications.

Input devices — keyboard, mouse, character-recognition devices, microphone, video cameras, scanners, Kimball tags, bar codes, magnetic ink character readers (MICRs), etc. — are able to convert information in any form (data, text, speech, or image) into binary pulses by computer.

Interface is a general term used to describe a shared boundary between two related devices or components defined for the purpose of specifying the type and form of signals passing between them.

Internet If not capitalized the term refers to any collection of distinct networks interconnected by routes that allow the networks to operate as one internet work. However, when capitalized the term refers to the worldwide network of networks connected to each other using an IP (Internet Protocol) suite. Intelligent networks provide the bridge between telecommunications and computing. Transmitted information is manipulated by the communication networks, for example 0800 telephone numbers.

Interpreter

- A unit that reads a punched card and prints on the card the characters coded by the hole patterns.

- A machine-code program used to translate and carry out the instructions of a high-level program one by one according to need. Interpreting rather than compiling allows interactive programming and does not produce object code. On the other hand, the interpreter must remain in store all the time the high-level program runs. An interpretive program language is one where the design makes interpretation rather than compilation essential (or at least, highly suitable).

Intranets are secure, corporate-wide, private networks, like an internal web site, that are incorporated inside a company's network. Intranet — the application of Internet technologies such as E-mail, discussion groups and web sites for internal/private information systems — made its presence felt in early 1996 as an integral part of most leading IT supplier product strategy. In 1997 IBM, Lotus Netscape, Microsoft, Novell and Oracle positioned intranet as the key component of their business and strategy. General Electric was saving $250 000 in printing costs simply by publishing company information directly on its web. The natural progression of intranet from a publishing vehicle of information at a departmental and group level, to being the vehicle for information sharing, collaborative interchange and giving significant improvements in workgroup productivity is where intranet will give real business gains with a high return on investment (ROI).

ISBN (International Standard Book Number) each book published is allocated its own unique number (ISBN) consisting of ten digits for use in aiding the location of a book within a library.

J

Java is a programming language developed by Sun Microsystems that can be read and translated, to run on any platform. Java allows the creation of interactive multi-media applications delivered over the network. Internet applications can use sound and 3D graphics yet be platform-independent running on UNIX, Windows 95 and 98, Windows NT and Mac.

Jig A device that holds and locates a workpiece, but also guides controls, or limits one or more cutting tools.

Job A specific piece of work to be input to a computer. Each job normally requires a number of runs.

Jump is a change in sequence of the execution of the program instruction, altering the program. In printing the term means to carry over a portion of a newspaper or periodical feature from one page to another.

K

Kinematics (from the Greek word meaning movement) is a process for plotting or animating the motion of parts in a machine or a structure under design on a system. Kinematic synthesis is primarily concerned with the design of kinematic systems capable of achieving particular required motions.

L

Language A set of rules for creating instructions that either can be understood by a computer or can be translated into something that can be understood by a computer such as a *low-level programming* language in which statements translate on a one-for-one basis. *Assembly language* sends instructions to the computer at a very fundamental level. Programs written in assembler are more difficult to write but generally run faster. They are usually used by experienced programmers. Machine language is the only language directly understood by computers, and is the lowest form. *Symbolic language* is 'human-oriented programming language' prepared in coding other than the specific machine language, and thus must be translated by compiling, assembly or other means. Interpretive language allows a source program to be translated by an interpreter for use in a computer. The interpreter translates the interpretive-language source program into machine code, and instead of producing an object program, lets the program be immediately operated on by the computer. *Interpretive language* programs are used for solving difficult problems and for running short programs. *Source language* is a symbolic language comprising statements and formulas used in computer processing. It is translated into object language (object code) by an assembler or compiler for execution by a computer. While the compiler translates an entire high-level language into a language the compiler can understand, the interpreter translates a high-level language into a machine language one statement at a time.

Layer is the term used for structuring a large quantity of data (e.g. CAD drawings containing a million characters of data) in a computer. These categories in CAD drawings are analogous to sheets of tracing paper in overlay drafting. Each layer can be owned by a different designer. They can all be turned on for co-ordinating the design or selectively turned off to provide only the data relevant, e.g. to one sub-contractor or other work sections. Layers may also be assigned to drawings elements such as text dimensions or hatching where the data displayed on computer screen need to be reduced. A British Standard BS 1192:P5 was published in 1990. It recommends organizing layers according to a widely used classification system such as CI/SfB Table 1 and Common Arrangement. The number of layers used on any project should be the minimum necessary for exchanges agreed by all concerned. The structured data can then be passed on disk or by telephone using DXF translators available with most CAD systems. The AutoCAD user Group (0483 303322) has proposed a layer naming convention using CI/SfB, allowing eight fields of data of which the first two, the discipline owning the data and CI/SfB number are mandatory. The others allow for graphical elements, pen-type, floor-level, new or existing work, scale and phasing.

Linux Unix-based system represents an alternative to Windows platform.

Local Area Network (LAN) A network providing communications between users within a defined locality. Such a network may be independent of public services. Compare with CLAN (Cordless Local Area Network).

Lockheed is the world's largest database host offering on-line access to over one hundred databases covering a broad spectrum of subjects.

M

Macro A recorded (saved on disk) set of keys (a stored sequence of keystrokes): function, menu, dialogue box entries, file names and so on which can be replayed by running the macro (usually by pressing Ctrl or Alt and another). When recorded the macro is given a name — the least number of characters which is capable of identifying it from, all other macros — and to repeat the complicated function just the macro name is selected.

Main storage is that part of a computer which receives data in the form of binary digits and stores it for future use. Storage and memory are synonymous. Main storage usually has a fast access time (between 100 nanoseconds and 1 microsecond) but because it is relatively costly its size is also restricted. However, memory size is also restricted by the addressing capability of the computer, e.g. on an 8 bit microcomputer it is 64k bytes. On a larger computer which has a 24 bit address field, it is 16 megabytes.

Mechatronics The integration of mechanics and electronics in the CAD/CAM context. This would be an integrated mechanical CAD(MCAD) and electronic design automation (EDA) tool used in the production of a part with components associated with the two technologies. An example would be a mechanical device of which a printed circuit board was part of the overall apparatus.

Memory Type of data storage for a computer system which is of two types

(a) main memory (or primary data storage) which holds the program(s) and data actually being used while the computer is running and is usually lost when it is switched off

(b) secondary (or external data storage or backing store) refers to the extended memory of the computer using a medium such as a disk, cassette and so on and which is retained when the computer is switched off.

Modelling refers to descriptions using data stored in computer memory. Surface modelling is used when 3D geometrics are required for the exterior shell of a product. Solid modelling enables a designer to sculpt a complex object rather than construct it with a series of lines. Four techniques in sculpting are union, intersect, difference and sweeping. Two basic methods of solid modelling comprise constructive solid geometry (CSG) in which complex shapes are made by adding or subtracting simple shapes in building block fashion and *B*oundary *REP*resentation (B-REP) in which each face edge and vertex boundary and object is explicitly defined.

Modem (from MOdulator DEModulator) A device that both modulates and demodulates transmitted messages. Modem is a hardware device that permits computers and terminals to communicate with each other using analog circuits such as telephone lines. The modem's modulator translates the digital signals into analog signals that can be transmitted over a telephone line. The modem's demodulator converts analog signals into digital signals for the computer's use.

Mouse is a small device that is pushed about on a table-top, the relative movement being fed to the computer. It can therefore be used to move an arrow around the screen or for building up drawings (though this has been described as being like drawing with a house brick). They operate by either of two principles

   (*a*) most simply feature a rubber-coated ball in their base which is rotated by movement over a contact surface, rollers then convey input information back to the computer

   (*b*) other mice have no moving parts but instead trace operator actions via firing infrared beams back and forth at a special mouse mat etched with fine reflective lines.

The mouse can arouse quite strong feelings of warmth or hatred. The haters prefer trackballs or trackwheels.

Multi-media refers to mixing text, images, sound and video into computer-based presentations. Typical uses in building construction might be to sell a development before it is completed by using 3D computer animations. A multi-media PC equipped with a sound card and/or audio speakers is used to access CD-ROM discs which contain data just like a floppy disk but consists of multi-media sound and vision.

N

Network

- A series of interconnected points.

- In communications, a system of interconnected communication facilities. An interconnection of computer-based devices and systems provides information services and transmissions covering a number of locations, e.g. LAN (Local Area Network). CLAN (Cordless Local Area Network) allows computers to communicate with each other and with peripherals (printers etc.) without the need for cabling. Most of these systems work with radio waves and have a small antenna plugged into the back of desktop PC creating the link to a control unit on (or in) the ceiling. Other products use infra red and Infra Red Data Association (IRDA) has agreed global standards to be adopted. Peer-to-peer networks allow users to automatically share the same document, work on it simultaneously and then all depart with an up-to-date version.

- In data structures, a structure in which any node may be connected to any other node.

Network computer (NetPC, NC) refers to a slimmed down low-cost computer created primarily for browsing web pages, handling E-mail and doing basic computing tasks. It is cheaper because many of its operations are performed through the server on a corporate network or a web server. Little in the way of applications is held locally.

Bill Gates, founder of Microsoft which have teamed with Intel to back the NetPC system says

The most important new hardware approach is the NetPC and it is very important that we are clear that the NetPC is just a form of the PC; it is totally and absolutely compatible. Its nice and small, its very attractively priced, its got a sealed case design. But these machines are fully capable of running Windows 95 and NT. We have made it very easy to boot, even across the network. And so there is no compatibility issue with these machines. You can mix and match them from different vendors or with PCs of other types.[2]

Larry Ellison of Oracle Corp. notes

Only a low-cost and convenient device like the NC will enable worldwide plans to spread literacy, education, and communication through computer access. Network computers, coupled with the Internet global network, can create a network economy and, yes, a network community. The Network computer assumes the existence of a network. The complexity is moved into the network from the desktop. The NC computer requires only two power cords, one for information from the network connection and one for power.[3]

O

Object-oriented programming refers to a software development that treats data as a complete entity or grouping of objects rather than as a simple 2D table of rows and columns.

On-line

- Relating to equipment, devices or systems under direct control of the central processing unit.

- Operation where input data is fed directly from measuring devices into the CPU or MCU. Results are obtained in real time; i.e. computations and operations are based on current values of operating data and answers are obtained in time to permit effective control action. On-line can also mean the operation of peripheral equipment in conjunction with the central processor of a computer system.

- In teleprocessing, a system in which the input data enters the computer directly from the point or origin and/or in which data is transmitted directly to where it is used. Therefore a device is on-line if it is directly connected to and hence under the control of the central processor.

Open Systems Interconnection (OSI) is a standardized set of procedures for the exchange of information between computers, terminals, networks and so on.

Optical fibre is a very thin flexible fibre of pure glass used to carry as much as a thousand times the information possible with traditional copper wire. Two main light sources for optical fibres are lasers and light emitting diodes (LEDs)

Outsourcing is when organizations transfer the ownership and/or management responsibility of their information systems to an outside supplier, but pay a fee to do so.

P

Password is a word, phrase or group of letters or digits that enable a user to gain access to a location, system, computer or data.

PDM (Product Data Management) is defined as a technology that is used to manage

(*a*) all product-related information, i.e. any information that describes the product including things such as part information, configurations, documents, CAD files, authorization information, etc.

(*b*) all product-related processes, i.e. both definition and management of the processes including authorization and distribution information.

This definition provides a broad view of the scope of PDM and its application to industry. The capabilities provided as part of PDM systems fall into five primary functional areas (and number of utility-oriented areas): data vault and document management, process/work flow management, classification and retrieval, and finally program project management. (Refer to *PDM Technology Guide*.)[4]

Performance specification A performance specification seeks to spell out design intentions, quality and durability expected, levels of thermal performance, sound insulation and fire protection requirements, provision for piping and conduits, foreseeable interfacing problems at package boundaries, the nature and length and bonding requirements for warranties or guarantees, etc.

Peripheral A device under the control of a central processor which performs an auxiliary action in a system, e.g. input, output, backing store.

Pixel (contraction of picture element) refers to the smallest unit of resolution on a raster-scan display. On a computer system, a pixel is the smallest portion of the screen that can have its light characteristics converted to computer-readable expressions of electric current. A matrix of dots stored in the computer memory defines an image for computer graphics. This matrix is converted into a video signal by a DAC (Digital Analogue Convertor) before display on a monitor.

Phoneme the smallest element of spoken language which distinguishes one utterance from another, e.g. the word 'bit' consists of 3 phonemes -/b/, /i/ and /t/.

Plotters Output devices like printers but can produce maps, graphs, plans and pictures unlike printers which produce text or tables. Historically plotters have been vector devices in which information sent to all plotter types is sent in vector format and if necessary converted back to raster format again. There are also three main types of plotters

(*a*) Pen plotters in which one or more pens (pencils, fibre, point pens, drafting ink pens, etc.) mechanically move to draw the plot.

(*b*) Direct thermal plotters (raster devices) which work by heating a specially treated paper — when the paper reaches a certain temperature it turns black.

(*c*) Electrostatic plotters (the fastest and most expensive plotters) work on a similar electrostatic principle to photocopiers and laser printers — a writing head lays an electrostatic charge on the paper and toner particles are attracted to the charged areas.

Plug and play A hardware and software design that is supposed to automatically configure system resource settings. Before 'plug and play', if one wanted to add a piece of hardware to a system, one had to turn off the PC before installing the component.

Plug-ins Software extensions downloaded from the web let users interact with intelligent images.

POST (Power On Self Test) A procedure that a computer goes through when booting or performing routines necessary to get all components functioning properly before loading the operating system.

Printers (output devices) A classification of computer printers[5] shows characterizing printers which print one character at a time, line printers which print one line of print at a time and page printers which print a whole page at a time. For example

• a daisywheel printer is a type of slow, high quality printer that cannot produce graphics

• ink-jet or bubblejet printers spray ink on to the paper to create high-quality text and graphics output

• a laser printer is a fast, high-quality output device using Xerox technology to output computer text and graphics one page at a time.

Product modelling is an agreed information model that defines how information should be structured before it is passed to another party.

Programming refers to the process by which a computer is made to perform a specified task. It involves the creation of a formalized sequence of instructions (called programs — static entity) which can be recognized and implemented by the machine.

Protocol is a set of rules. In particular, a set of rules governing the way that information can flow in a system. Such rules cover syntax (the structure of commands and responses), semantics (the requests, responses and actions that users can perform), and timing (the sequence and ordering of events).

Q

Query In data communications, the process by which a master station asks a slave to identify itself and to give its status.

R

Radix In a radix numeration system, the total value of a numerical represented by a string of characters is the sum of each character multiplied by weight. The ratio of the weight of one digit to the preceding one is the radix for the number system used, and is always a positive integer. Thus in a hexadecimal system the radix is 16 and the number 123 has the decimal value of $291 = 1 \times 16 \times 16 + 2 \times 16 + 3$.

Random Access Memory (RAM) is an auxiliary memory device on which the programmer can directly access each separate data area without having to search through the whole data file. Dynamic RAM (DRAM) requires periodic refreshing (every 2 milliseconds) because of charge leaking from capacitors in the cell circuit (it is an inexpensive MOS random access memory). Static RAM (SRAM) does not require refreshing because of the cell circuit design (although it is still volatile, losing data on removal of power). SRAM is more expensive than DRAM, but is faster and requires less control circuitry. Compare with ROM a memory which cannot be modified or reprogrammed and typically is used for control and program execution.

Raster refers to the pattern of lines formed by the movement of an electron beam inside a screen or camera when scanning a picture. The horizontal scanning pattern of the electron gun in TV sets and computer monitors that use CRT (cathode-ray tube) display. (Contrast with vector line or 'calligraphic'. A vector display device that stores and displays data as line segments identified by the X,Y,Z co-ordinates of their end points.) A raster display device stores and displays data as horizontal rows of uniform grid or picture cells (pixels). A raster unit is the distance between two adjacent addressable points located horizontally or vertically on a CRT display.

Rendering is a term used to refer to turning CAD 3D model into a picture — this includes all the aspects that makes a good photo such as camera view, composition, light exposure, etc.

RGB (red/green/blue) Commonly used term to refer to the colour space, mixing system, or monitor in colour computer graphics. In RGB, a colour is defined as percentages of red, green and blue, with 0,0,0 equivalent to black and 100,100,100 equivalent to white. Shades of grey have equivalent percents of red, green and blue and 0 and 100 i.e. 25,25,25 is a dark grey. A RGB monitor is the type of CRT screen that produces colour images when a trio of red, green and blue electron guns focus on phosphor triads.

Robot is a programmable machine intended to perform a job originally conceived in terms of a human operator. There is a programmable multi-functional manipulator (a robot's arm, hand). A prosthetic robot is a mechanical device connected to the human body which provides a substitute for human arms or legs when their function is lost.

S

San serif type refers to any type face where the characters do not have 'serifs', small side strokes at the ends of the main strokes.

Scanning A process in which Trace software is used to import drawings into CAD programs or in which Scanner software is used to superimpose complex images on technical line drawings, for example scanning images of trees or brick walls and incorporating them in architectural designs. Label scanners are widely used in supermarkets and libraries to read bar codes by either a wand or by passing the label over a scanning window. Bar codes are usually a set of vertical lines of varying thickness or a set of concentric circles. Since bar codes can be produced by a computer printer they can also be used for stock-taking and even for more esoteric purposes such as recording money for parking meters.

Search engine sites Table 2 lists the top search engines available.

Semantics refers to

- that branch of the study of language concerned with meaning
- that branch of logic concerned with the truth values of relationships (in computing, truth tables represent these values)
- the study of symbols and how they relate to what they represent.

Semi-conductor Any substance or material with a conductivity midway between that of an insulator and a good conductor. The conductivity is sensitive to temperature, radiation and the presence of impurities. Such materials are used in the manufacture of transistors, diodes, photoelectric devices and solar cells.

Sensor A transducer or other device whose input is a physical phenomenon and whose output is a quantitative measure of that physical phenomenon.

Serial A data communication standard where the data is carried over three wires (send, receive and earth (ground)). Serial interface refers to the connection between PCs or between PC and auxiliary device, e.g. printer, modem, which transfers information one bit at a time. A serial printer receives its input through a serial interface.

Server The computer system that contains information such as electronic mail, data information or text file.

Shape refers to a set of line segments.

Shell

- User interface software: a program that accepts user commands, interprets them, and passes them for execution.

- In general any program that protects a user at a higher level from the problems of interacting with software at lower level. This usage is very common in work with machine intelligence.

- A type of data-sorting program, strictly 'shellsort', the name given by the inventor, Donald Shell (1959). This variant of an insertion sort allows data items or records to jump a large distance in the set rather than move step by step.

Software Programs, languages, procedures, rules and associated documentation used in the operation of a data-processing system. Software package is a set of programs for a specific purpose and often sold independently of hardware.

Speech is digitized and decomposed into units of sound (phenomes) which are matched against the phenomes patterns of words in standard English (or another language). The systems do not rely only on vocabularies since words which sound identical may be spelt completely differently and such vagaries can only be determined by reference to a language model which understands the context in which, say four and for are defined. DragonDictate and IBM Voice-Type Dictation are speaker-dependent, i.e. designed to be framed to specifier user's voice patterns. The third component therefore is the individual speaker model which is created when we start using one of the dictation systems, and then refined continually as we work with them. Overall IBM's language model and user model is more complex than Dragon's, while Dragon's implementation is more customizable. The Dragon system recognizes words on a word-by-word basis while the IBM system takes far more account of the context. This effect can be seen as one dictates into the system. Words are continually parsed up to three words behind and strange meaningless phrases appear on screen before the actual words you said appear. It only takes seconds but it is the reason IBM uses the Dictation Window approach, and it results in very accurate and rapid dictation. With DragonDictate you are dictating into a small buffer which runs as a keyboard TSR and there is limited scope for the kind of processing being performed in the IBM Dictation Window.

Speech Recognition Technology Speech recognition systems may be 'speaker-dependent' or 'speaker-independent' systems which have three components: the vocabulary or dictionary, a language modeller and the speech model.

Spreadsheet This term refers to a sheet of paper which is spread out and then filled with a table or matrix of rows and columns. Spreadsheets by computer avoid the long-winded calculations by hand which are notoriously prone to calculation errors especially if repeated.

Table 2

	URLs	Date of launch	Spider program	Description
AltaVista	www.altavista.digital.com/	December 1995	Scooter	Funded by DEC (Digital Equipment Corporation). Select word search. Tends to produce many obscure references, especially in unskilled hands, but Live Topics search feature may make things easier.
Excite	www.excite.com/	October 1995	Architect Spider	Arranges results in order of relevance, and if it finds what you want, you can click to get 'more like this'. Enclosing phrases in double quotation marks to look for the complete phrase rather than just the occurrence of the individual words is a good tip.
Hotbot	www.hotbot.com/	May 1996	Slurp the Web Hound	Excellent index, and the menu system makes it the easiest place to do complicated Boolean searches. Emphasizes product sites. However, the colour schemes sometimes seem to be designed for people wearing sunglasses.
Infoseek	www.infoseek.com/	February 1995	Spider Winder	Powerful engine that searches the whole web or focuses on nine major topic sections. Allows one to search the web for a word or phrase. A hybrid service that offers 15 'intelligent channels' and a search engine that can accept plain text queries (e.g. Who wrote Romeo and Juliet?) even if it does not always find the answers. Also provides related sites.
Lycos	www.lycos.com/	May 1994	T-Rex	Extensive index of documents, including words in title, headings, sub-headings and hyperlinks. Offers 'channels' called Web Guides, and includes a directory of sites rated as being the top five percent.
Open Text	index.opentext.net/	1991		Provides knowledge and management solutions for Global 2000 companies. Also delivered one of the first search engines on Netscape Communications' home page and provided the original back-end technology for Yahoo.
UK Index	www.ukindex.co.uk	1995	none	Database of almost exclusively UK sites with vetted sections. The leading source of information on the Internet in the UK. Set up in response to the problems of using other systems, namely searching the actual contents of the pages for useful text (at its worst the page may consist entirely of graphics!) and results always had a strong US bias, a situation hardly ideal for UK users. UK index assigns the pages to broad categories. The idea behind this is that with a free form system it may be far from obvious what words are used to describe a resource, but if categories are there to pick from it should be a lot clearer. The final touch was the mailing list — a way for people to keep up to date with the subjects they choose.
Yahoo	www.yahoo.com/	After April 1984. Created at Stanford University by D. Filo and J. Yang.	none	Yahoo stands for Yet Another Hierarchically Organized (or Officious) Oracle. A database of World Wide Web pages organized by subject and based on a catalogue of indexed resources. For years the only search engine on the web. By far the Web's largest hierarchical directory and lists more than 500 000 sites in 14 main categories. It links to other search engines to provide Internet searches, with AltaVista being the first choice.

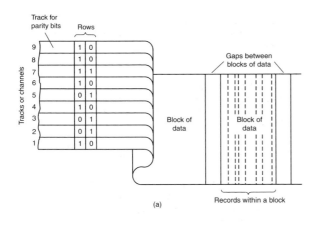

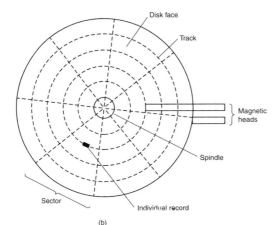

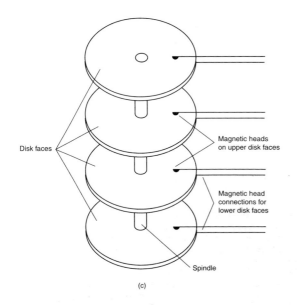

Fig. 1. Secondary computer storage:[6] **(a) data storage on magnetic tape, (b) data storage on magnetic disk, (c) disk stack with magnetic heads**

Storage devices (disks, diskettes, tapes, drums, etc.) Figure 1 shows some of the main means of secondary computer storage as opposed to main computer storage.[6] Disks are flat plastic platters that resemble phonograph records used to store information. There are flexible floppy disks with 5 1/4 inch, 8 inch diameter, 'hard' disks. Magnetic tape can only retrieve stored information sequentially, i.e. by researching the tape from beginning to end. Magnetic drum is a right circular cylinder with a magnetic surface on which information can be stored by selective magnetization of parts of the curved surface.

Suite of office programs Table 3 provides a comparison of popular office programs.

Table 3

	Manufacturer	Description
Corel office	Corel Corporation	WordPerfect word processing, Quattro Pro database, Presentations and a personal information manager and document and packaging tool.
Microsoft office	Microsoft Corporation	Word word processing, Excel spreadsheet, Powerpoint Presentation, Mail, Office assistant, Access database allows web-site hyperlinks.
Lotus SmartSuite	Lotus Corporation	Word Pro word processing, 1-2-3 spreadsheet, Freelance, Approach database, time management system and screen action recorder.

Syntax Rules for the exact arrangement of the characters and words which make up a programming language. Compare with *Semantics*.

System files Small disk files containing the first software codes a computer reads from disk once it has been turned on and booted to load the rest of an operating system. Ordinarily, system files C10.SYS and MSDOS.SYS on DOS and Windows respectively) are not seen in a listing of files on the disk. Another system file COMMAND.COM in DOS contains the operating system's basic functions, e.g. displaying a list of files (directory). System files may also include CONFIG-SYS, which makes some initial settings of hardware. AUTOEXEC.BAT is a collection of commands that are executed when all other boot functions are finished. In Windows 95 and 98, the Registry, which consists of the two hidden files USER.DAT and SYSTEM.DAT, is also necessary for Windows to run.

T

Telecommunications Communications over distance using electronic means; types of telecommunications channels include twisted-pair telephone lines, coaxial cable, microwave, satellite, and fibre optic cable.

Teleconferencing is

- the use of telecommunications systems by groups of three or more people at two or more locations for the purpose of conferring with one another

- two-way communication between two or more groups, or three or more individuals remote from each other, using a telecommunications medium

- interactive group communication through an electronic medium.

Teleworking The practice of working from home using telephone, E-mail, fax and video to communicate with business associates.

Texture mapping is the application of a 2D image to a 3D object. To perform texture mapping, there needs to be definition of the texture. This is typically a 2D array of pixels stored in a rectangular organization. The data contained in these pixels can be red, green, blue and alpha components of the texture. A description of a 3D primitive or geometry to which the texture is to be mapped must also be provided. Then a mapping from the texture space to the 3D geometry space must be defined. It is important to note that the texture destination is not just a description of wood, plastic or steel but is actually data that can represent different types of information.

Trackballs/Trackwheels are small devices on which a small ball or wheel can be moved with one's fingertips. This movement gives directions to the computer. Trackballs can be more difficult to control than a mouse. However, users claim that they are faster, more responsive, more sophisticated, and much more accurate than the mouse. They do not require one to be ambidextrous in use and do not need valuable desk space. They are also more expensive than a mouse.

Transducers At the most general level, it is a device for converting energy from one form to another. In IT this is an input, or output, device designed to convert signals from one medium to another, e.g. a loudspeaker is a transducer which converts electrical signals into acoustic signals.

Transputers refers to the processor with serial links which allows communication with other transputers. This is the basis for parallel computers. Parallel processing architecture has been made possible by a British development, the Inmos Transputer — effectively a 'building block'. While it contains its own memory and processing elements, it also features unique serial links which allow it to communicate with other transputers. A matrix of transputers can be created with each one solving a small part of a complex task. The addition of extra transputers to a system incrementally adds full power of each unit to the overall performance.

TSR (Terminate-Stay-Resident) program remains in the PC's memory and can only be re-run by pressing a selected key combination (hot key). When a normal, non TSR program is ended, it is removed from memory.

Turing machine A mathematical model device invented by Alan Turing that reads data from tape, moves the tape zero or one position forward, or backward, writes to tape and changes one of its internal states.

Turnkey system A complete computer system designed for a specific user or a system in which the user needs only to switch on the particular system. The prime contractor accepts full responsibility for system design, installation, supply of hardware, software and documentation.

Typesetting The putting of text into typeset form on a medium, usually photographic film or paper, suitable for making printing plates. Consider computer-assisted typesetting, phototypesetting and later developments.

U

URL (Universal Resource Locator) refers to the address assigned to every web page. For example CAENET's URL is *http://www.penton.com/cae/* with each component defined as follows.

- 'http' identifies the location as one that used HTML (hypertext mark-up language). If the part before the slashes is 'ftp', then the location uses file transfer protocol.

- '://' alerts the browser that the next words will be the actual URL, which is broken-up by periods. Each period is usually referred to as a dot.

- 'www' identifies the location as part of the World Wide Web.

- 'Penton' is the domain name. This must be registered with Network Solutions, a company which has exclusive authority to register domain names under an agreement with the National Science Foundation.

- 'com' is the top-level domain name indicating the purpose of the sponsors of the site in this case worldwide 'commercial' entities, individuals or companies. Other top-level names include: 'ac.uk' for a British academic organization; '.edu' is assigned to higher-level educational establishments, colleges, universities, etc.; '.gov' is reserved for American government agencies and organizations, and similar bodies in other countries when followed by the relevant country code, e.g. '.gov.uk' for the United Kingdom; '.mod.uk' for the British Ministry of Defence websites; '.net' for organizations that are part of the Internet infrastructure (such as Internet service providers); '.nhs.uk' for the National Health Service websites; '.org' was originally reserved for non-profit making organizations (charities, political bodies, professional institutions, trade unions, etc.) but is now issued to some commercial enterprises; '.sch.uk' is the British school domain.

- 'cae' relates to a specific page at the site.

User Interfaces (GUIs, CLIs, etc.) The portion of the system software that displays messages to the user and handles the user's commands. A Graphical User Interface (GUI pronounced 'gooey') refers to a form of user interface employing icons to represent files and data, and a pointer to highlight options and commands in order to make these more user friendly. They make it easier for users to breakdown (or decompose) computer operations into meaningful sub-tasks, and permit a more coherent understanding of the system.[7] The most common GUI is Microsoft Windows. However, GUIs require greater processing power than CLIs. Command Line Interfaces (CLIs) require users to operate their computers by entering syntactically (appropriate format) exacting lines of commands through a keyboard. And because mistakes in entries when typing instructions are rarely tolerated CLIs alienate many potential users, e.g. in MS-DOS the CLI to copy information from one floppy disk to another is COPY A:\DOCS\LETTER.TXT B:\DOCS

Utility programs Programs supplied for common routine tasks, e.g. copying files.

V

Validation A manual or automatic check on input data for correctness against set criteria, e.g. format, ranges, etc.

Vapourware refers to computer products promised by manufacturers but that never actually appear.

Video pertains to visual images produced or transmitted by a television system. Videotex is a term proposed for use at the international level of Viewdata. The term is also used generically to cover teletext, a broadcast Videotex service as well as Viewdata, a wired Videotex service.

Videoconferencing refers to face-to-face meeting of people in different locations using cameras to transmit live pictures over telephone lines. More advanced systems provide shared work tools such as digital white boards and shared screens where users at both ends can annotate text or drawings during the video conference.

Virtual Reality (VR) The term refers to the artificial environment created with the medium of cyberspace — the virtual medium of computers in which software operates, which is also created when computers are linked across networks. Cyberspace is the electronic domain of pure information. VR represents the ultimate in graphical user interfaces. Therefore VR is the ability to explore a computer-generated world by actually being in it so that instead of looking at a screen you are enclosed in a 3D graphic universe where you can affect what happens to the virtual world as you do the real one. This definition covers a myriad of techniques and approaches aspects of which are in use. The two types of virtual reality are immersive VR — use of head-sets and in some cases full body armour alongside a proprietary parallel computer is regarded as true virtual reality — and non-immersive VR — just a step on from accepted visualization techniques. Immersion VR manipulates a 3D CAD model as though it were cardboard. By wearing data gloves, a computer with parallel processing generates an image of hands which follow the users' movements so that a wall can be moved or pieces of the model can be picked up. The underlying database is automatically updated. The head-mounted display contains two high-definition LCDs which project a stereoscopic image to each eye usually at a screen resolution of 360 x 240. There is also a head-tracking device used to correlate head movements with the display.

VRML (Virtual Reality Modelling Language) lets browsers view and interact with 3D models.

W

Wildcard refers to a symbol such as '*' which when used in a file name, represents any character or group of characters at that position in a name (e.g. DEL A:*.* i.e. delete all files in the disk drive A).

Window In computer graphics, the term means a bounded area within a display image which contains a scissored subset of the displayable data.

Wire-framed graphics is a CAD technique for displaying a 3D object on a CRT screen as a series of lines outlining its surface.

Wraparound is the effect of positioning a display item so that it extends beyond the device space boundary and a portion appears on the opposite side of the display surface.

Wrl stands for world, which is the extension that ends the name VRML model.

WWW (world wide web) is a hyper-linked, browsing multi-media environment.

WYSIWYG (What you see is what you get) pronounced 'wizzy wig' — a graphical user interface meaning what you see on the screen is an accurate representation of what you get.

X

Xerography refers to a process or technique in printing that first places an electrostatic charge on a plate and then an image is projected on to a plate causing the charge to dissipate in the illuminated areas, thus allowing an applied coating of resinous powder to adhere only to the charged (dark) areas. The powder is then transferred to paper and is fixed by heat.

XMS Memory refers to memory (in a 286, 386 or 486 computer) that has been configured as extended memory by a memory management program — extended memory specification.

Xon/Xoff flow control. Special characters sent during serial communication to stop or start the sender.

XYZ space A 3D co-ordinate system based on the 1931 CIE (Commission Internationale de L'Echairage) chromaticity which plots X, Y and Z as the tristiumlus values of a colour.

Y

Yaw

- The angular displacement of a moving body around an axis which is perpendicular to the line of motion and to the top side of the body

- In robotics, the rotation (especially of the hand) in a horizontal plane when the arm is extended horizontally.

Y2K bug (Year 2000 bug, also known as the Millennium bug). This is the mistake in the BIOS (Basic Input Output System) program in a computer which means that at midnight on 31 December 1999 the date on the machine is set back to an incorrect date. For the majority of affected machines, this date is 1 January 1900. When the computer is next turned on, the Operating System (DOS, Windows 95, 98 or NT, etc.) will object and set the date forward to 4 January 1980, the earliest date with which the computer can cope. The following day, if the date has not been corrected, the BIOS will think it is 2 January 1900 but DOS (or Windows) will object once more and set the date to 4 January 1980 again. Some BIOS programs, however, compound the problem further by setting the year back to that in which their program was written, and refuse every proper increment thereafter. This cycle will repeat, unless the problem is addressed, for the next 80 years or so. Hardly anyone was aware of the impending problem prior to 1996, so any computer purchased earlier than then is likely to be affected. This threat created by the IT industry exists for all organizations and businesses and the fall out may take years to rectify. Y2K compliance involves a testing process, which includes checks to see if the BIOS recognizes that the year 2000 is actually a leap year.

Z

Zero point The origin of a co-ordinate system.

Zoom In computer graphics, continuously scaling the elements of display image to more clearly perceive and manipulate details not readily perceived in the previous view.

References

1. Smith A. Computers manage the paperwork. *Building*, Oct. 1993, 30–31.

2. Wheelwright G. Place your bets for the Bill v. Larry showdown. *The Times*, 1 Oct., 1997.

3. *ibid.*, 1997.

4. Miller E. Managing engineering data: PDM Today. *Computer-aided engineering*. Penton Publications, Feb., 1995, 32–40.

5. Deeson E. *Collins Dictionary of Information Technology*. HarperCollins Publishers, Glasgow, 1991.

6. Redmill F. *The computer primer*. Addison-Wesley Publishing Company, Wokingham, 1987.

Acronyms

ADC	Analog(ue) to Digital Conversion		DXF	Drawing Exchange Format
AEC	Architectural—Engineering Computing		e-commerce	electronic commerce
AOL	America Online		EDMS	Engineering Document Management System
ASCII	American Standard Code for Information Interchange		EDVAC	Electronic Discrete Variable Automatic Computer
BBS	Bulletin Board Service		FAQ	Frequently Asked Questions
BIOS	Basic Input/Output System		ftp	file transfer protocol
bps	bits per second		GIF	Graphics Interchange Format
BSA	British Software Alliance		GPS	Global Positioning System
BSI	British Standards Institution		HCI	Human—Computer Interface
CAD/CAM	Computer-Aided Design/Computer-Aided Manufacturing		HTML	Hyper Text Mark-up Language
CD-ROM	Compact Disk Read Only Memory		IP	Internet Provider (sometimes Internet Protocol)
CISC	Complex Instruction Set Computing		ISA	Industry Standard Architecture
CNC	Computer Numerically Controlled		ISDN	Integrated Service Digital Network
CODASYL	COnference for DAta System Languages		K	(1024) measure of computer storage
CRT	Cathode Ray Tube		LAN	Local Area Network
DAC	Digital Analog Converter		Laser	light amplification by stimulated emission of radiation
DARPA	US Defense Advanced Research Projects Agency		LCD	Liquid Crystal Display
DAT	Digital Audio Tape		LSI	Large Scale Integration
DBMS	Database Management System		MAT	Machine-Aided Translation
DIN	Deutsche Industrie Norm		MIDI	Musical Instrument Digital Interface
DIP	Document Image Processing		MIME	Multipurpose Internet Mail Extensions
DMA	Direct Memory Access		MIS	Management Information System
DPI	Dots per inch		MPC	Multimedia Personal Computer
DTP	Desk Top Publishing		MPEG	Motion Pictures Expert Group
DVD	Digital Versatile Disk (previously digital video disk)		MS DOS	Microsoft Disk Operating System

NAP	Network Access Point
NBS	National Building Specification
NLQ	Near Letter Quality
OCR	Optical Character Recognition
OLE	Object Linking and Embedding
OS	Operating System
PABX	Private Automatic Branch Exchange
PBX	Private Branch Exchange
PCB	Printed Circuit Board
PCI	Peripheral Component Interconnect
PCMCIA	Personal Computer Memory Card Interface Association
PDA	Personal Digital Assistant
PDL	Page Description Language
PDM	Product Data Management
PMT	Photomechanical Transfer
POST	Power On Self Test
QIC	Quarter-Inch Cartridge
RAM	Random Access Memory
RIP	Raster Image Transfer
RISC	Reduced Instruction Set Computing
ROM	Read Only Memory

RSI	Repetitive Strain Injury
SMPT	Simple Mail Transfer Protocol
SOHO	Small Office/Home Office
STD	Subscriber Trunk Dialling
TCP/IP	Transmission Control Protocol/Internet Protocol
TIFF	Transfer Image File Format
TSR	Terminate Stay Resident
UDP	User Database Protocol
URL	Universal Resource Locator
VDT	Visual Display Terminal
VDU	Visual Display Unit
VLSI	Very Large Scale Integration
VR	Virtual Reality (Immersive or non-immersive)
VRAM	Video Random Access Memory
VRML	Virtual Reality Mark-up Language
WAN	Wide Area Network
WIMP	Window, Icon, Mouse and Pull-down menus
WORM	Write Once, Read Many
WP	Word Processing
WWW	World Wide Web
WYSIWYG	What You See Is What You Get